BLUE SCARS

BLUE SCARS
The Badge of a Miner

The caretaker of memories and trustee of the past

By Chopper Davies

AuthorHouse™
1663 Liberty Drive
Bloomington, IN 47403
www.authorhouse.com
Phone: 1-800-839-8640

© 2011 by Chopper Davies. All rights reserved.

No part of this book may be reproduced, stored in a retrieval system, or transmitted by any means without the written permission of the author.

First published by AuthorHouse   07/22/2011

ISBN: 978-1-4567-8373-0 (sc)
ISBN: 978-1-4567-8603-8 (ebk)

Printed in the United States of America

Any people depicted in stock imagery provided by Thinkstock are models, and such images are being used for illustrative purposes only.
Certain stock imagery © Thinkstock.

This book is printed on acid-free paper.

Because of the dynamic nature of the Internet, any web addresses or links contained in this book may have changed since publication and may no longer be valid. The views expressed in this work are solely those of the author and do not necessarily reflect the views of the publisher, and the publisher hereby disclaims any responsibility for them.

Foreword

This book is dedicated to my wife Tilly who made it possible to fulfil a dream to which in time may prove fruitful. No better partner a man could ever wish for. This is my tribute to a woman who without her loyalty and support I could never accomplish my unfulfilled ambition to acquire some form of status as an artist. And her inspiration played a big role in my attempt to further my ambition by nurturing that spark into a flame, and was always the one that made the visits of many celebrities to my home feel the warmth of her generosity in dealing with unexpected guests. But I have lost all this by her passing, but owe it to her memory to try and carry on with my work. This is to pay tribute to the force behind my work who was always in control; that was my wife Tilly. What more can anyone wish for and it is in her spirit and my love of art that can rekindle the driving force that was once there. So if this social document is ever published, it is an insight in the life of coal miner. And also a tribute to my mate who lost their lives at SIX BELLS. This is the only opportunity I will have to thank those men of character who played a big part in my life in the darkened city beneath our very feet.

John (Chopper) Davies
2005

Blue Scars

My name is John (Chopper) Davies born and bred in Nantyglo. This is a true story, from the heart, pictorial reminiscences of a miner's life in the dark world of the pits, a life where there are few traces left in the South Wales Valleys. There is no plagiarism in this story that I tell, and I do not use half hidden symbolised meanings, nor do I glamorise my life down the pits. My story is to draw attention and provoke thought to future generations, of their heritage, having the experience of forty two years behind me and having played an active role in the pits, gave me the advantage and determination to write this social document, highlighting our heritage, and also a tribute to the earlier pioneers of my dark world. There are no sentiments in the telling of my story because lurking around in the recesses of a darkened coal mine are those hidden dangers—namely gases, roof falls and many transport problems.

Most every street in these South Wales Valleys testifies to some form of tragedy that has befallen families. Their husbands, sons and friends that have left deep scars that are left in men and women in the pursuance of those black diamonds, which have laid the foundations for the prosperity of our valleys. There are moments of sadness that I myself have witnessed during my life in that dark city beneath our feet. Tales of heroism that go unseen in the darkness of the gloom of those passages of hidden perils. Many times I have heard the words of 'thanks butty' uttered by a grateful mate that works in close proximity to each other, and it creates a atmosphere of confidence knowing the relationship in that bond of camaraderie, among your fellow miners. This is why my story sets precedence for my individualistic reminisces of life that goes on in that labyrinth of passages beneath the earth surface as many miners can testify to.

Chopper Davies

The story is not particularly motivated, but it plays a large part in the telling of a struggle futile though it may be against the Government of the day, Thatcherism, and this is why I highlight the trials and tribulations that has befallen our great industry, but is also happens in successions of Government that opposed any attempt by unions to improve conditions and wages, that were the entitlement of their members, but that's another story to be told later on.

I made my life down the pits as agreeable as I could make it, after leaving the umbrella of school life to enter the adult life of toil and uncertainty of the miner's world, but not through choice but necessity by helping to ward off hardships that could develop in the household. My contribution was of a minute quantity, a mere pittance at the age of fourteen in 1936, but nevertheless was important to keep the wolf at bay and having the effect of self respect and a feeling of self importance, which was absent prior to becoming a pit boy, in which eventually became my destiny, and far from my umbrella of school days in which I was grateful for the teaching that I had in which I was not the best of pupils, in fact I was a 'duffer' and as dull as a sledge hammer!, not that you needed any special qualifications in respect of the early stages of mining life, and in my present stage was determined not to make a career of the pits, but then again one cannot foresee ahead of what the future holds for you, but in one respect unknown to me at the time that by doodling with oil and chalk on rusty sheets and rusty drams, or an occasional whitewashed wall when time permitted me to do so, in the twenty minutes food time, when that time came around, soon fostered my craving to draw, but quite often I was reprimanded by management for the embellishments added to my drawings by the mischievousness of passing workmen, which did nothing to improve the characteristic personality of the persons that I had created, but also it gave a cartoon quality to its effect, but by the managers point of view when defacing whitewashed walls, made it a sort of graffiti to which management could not afford to ignore because the whitewashed walls were there to enhance the illumination of the feeble light bulbs that tried to light their way through dust covered mesh. So, I was cautioned to curtail my enthusiasm for doodling in that sort of art form but I am afraid it did not deter me from my quest for the knowledge of how to progress from there, because to me words cannot express feelings but to me, my paintings captured the imagination and your own interpretation of what you are seeing rather than what you are listening to, but I was told by management to concentrate on something more constructive (they as good as told me to stop messing about I suppose), but there was an abundance of oil and chalk and sootiness that I could not ignore in my quest for drawing, so I carried on regardless as to what the outcome may be, but later on proved invaluable in safety values, but I could not foresee that far ahead—but that's another story, but, it was all the motivation that I needed.

My life as a school boy—needless to say I was not in any way the brightest and because of my physical stature was often bullied by the bigger lads, hence being absent from school on so many occasions at the first chance that I had to avoid such dealings with that part of my school days and that element, but I weathered the storm so to speak over the years and I was grateful for what teaching that I did have because I could read and write as good as anyone in my class at that time, but even then, my favourite subjects were writing and drawing and after those subjects I call it a day, but on my favourite subjects I was let out of school early for being the most attentive, but after general lessons I can call it a day and I make no excuse for it.

I carried on with my art work and was later encouraged by management, who by the way gave up on trying to discourage me, and persuaded me to paint safety posters and provide the necessary and essential scenarios to portray safety posters and their values associated with safety, but prior to my work on safety, meetings were held on the matter of safety attended by the unions and represented by a few officials, but did not include many ordinary workmen, so these verbally were not good enough to put safety values across to the workmen, but when my paintings were used as scenery for little plays, that were acted out at little local venues, it attracted more of an audience and small dramatised versions developed. It more or less was a novelty in its infancy but the seed was there to attract larger audiences, including the wives of the local community. I was asked by union and management alike to enlarge these paintings into backdrops for these plays, the murals were 8 feet by 8 feet and were sufficient to encourage budding actors to act. It was more of a giggle than anything else, but surprisingly turned serious as time progressed, because it was unusual as a form of entertainment and it had a surprising effect on workmen in the clubs. It also attracted their families who regularly attended these little one off shows and also gave the wives and children an insight into what conditions their families worked under, so these little versions of pit life grew and grew and because of the novelty of some local actors, they attracted a following till it came to pits competing against one another in exhibitions—all because of that bit of doodling with oil and chalk!

Blue Scars

I was proud of my accomplishments that started at the age of fourteen but it was the pits that gave me that opportunity to become a mining artist and this is my tribute to the works at Six Bells and Beynon Collieries that put me in a position to attain some status not only as an artist but also as a workman for forty-two years, and if the pits were open today and I were younger I would go back tomorrow because I miss the comradeship of such fine men, but alas, there is no hope of that becoming a reality. I exhibited my work at different venues throughout the valleys, but what I loved best was the visits to schools, teaching and explaining to children about their heritage, my time has not been wasted, because over the period of thirty-five years I have missed a lot of social life, by being dedicated to painting of which I collated over 400 works of art and not counting the countless hours of sometimes thirty to thirty-seven hours a week, so I still keep my mind active today to continue my work, but all this could not have been done without the encouragement of a fine and honest wife that I had—Tilly. She was my inspiration and without her I could not have done what I had set out to do. My wife played the part of a neglected wife most of the time because I was always shutting myself off in my spare room and only came down when my wife called me to my food times, to which my wife provided for me.

My hopes were not dampened by some of the setbacks I encountered. I still maintained my ambition of being an artist and was always encouraged by the comments of school teachers and pupils alike when exhibiting at their schools, lecturing children about their forgotten heritage and what inspired me most was the interest of the parents when collecting their children from school because they had no idea of the workings of a coal mine and the conditions men worked under at their tasks in that gloom. One lady came up to me on one occasion and said she was under the impression that there were locomotives and trucks that were used on the main lines and wondered how miners could fill those trucks in such a confined space as those small tunnels, but after looking at some of my paintings realised what a silly question she had asked. But I assured her that most people were of the same mind as her and after explaining the principle in which machinery worked and that an engine was on a stationary platform on which two revolving drums or large pulleys with ropes attached to haul the drums along the rail track, then I drew her attention to another picture with a further illustration of how the connection worked between engine and drums, that answered and satisfied her curiosity.

I enjoyed every minute in these classrooms answering questions from the children and teachers alike. Sometimes you do have an odd question, typical of some children and some questions you have some difficulty in answering, like the one about a painting of a pit shaft with a cage descending down the shaft into the darkness revealing the surrounding brickwork, glistening with the wetness and oiliness from the exposure to all kinds of weather but still revealed the original orange coloured brickwork which made one of children ask this question 'Sir' he said, 'How many bricks were used to make that hole in the ground?'. Typical of some children in their innocence but it gives you a sense of guilt when you know it was impossible answer. In some schools depending on ages I used to soften the harshness of pit life to some of my classes by telling them a story of some of my incidents while I worked down in the pit darkness and I told this story in most schools I have visited during my exhibitions which I knew the younger ones would be most attentive.

Blue Scars

I began by telling them as I sat down one day in the darkness of the pit the only illumination I had was the light from my Davy lamp (oil lamp) and was having my food. The children were becoming interested now and began to form a half circle in front of me and were wrapped up in their thoughts in anticipation of what I was going to tell them of my friendship with a little mouse in the dark tunnels. I captured their attention with a few white lies and by that time a picture was forming in their minds and needless to say I myself was getting deeper into my story, but prior to that I had manipulated a form of a mouse shape with my handkerchief to create a more subtle approach to the story. The story began with one day while sitting on an old log of wood, eating my food, because it was food time, I saw a little flurry of grey but it was gone in a flash—it was a little mouse picking up crumbs of bread that had fallen off my sandwich, but he was gone again in that quick movement that mouse have, and scuttled into the cracks of the stones. This went on for a few days while I was having food but as the days passed by he became bolder knowing that I wasn't going to hurt him and he stayed longer to nibble at the crumbs I had thrown down, but he kept his little red eyes on me, ready as ever to dart away, then he started to get even bolder and came close to my knees. I threw bits of cheese and he nibbled away at it and swished his long tail like a dog wagging his tail because he was pleased as he knew in his own little mind that I meant him no harm. I did not have a lot of time to mess with the mouse because I had to start to work again, so now the children were getting really absorbed with my story, but unfortunately the dinner bell went for their dinner break, but as soon as the dinner break ended they were back and could not get back fast enough to hear more about my little friend. S I told them I had not seen my friend for a few days now because I was working elsewhere, so after a few more days had gone by I returned to my old place of work and I was walking along the roadway when I hear a lot of squeaking—it was more than one mouse, so I threw little stones at them to clear them off but they would not go away, they seemed to be following me so I stood and looked at them and I could not believe my eyes, it was my little friend and he had come to show me his family and what amazed me—they were playing about with my friend, but after a few more squeaks they disappeared into the stone wall, but my little friend stayed a bit longer and looked at me, then he too disappeared and I knew then I would not see him again and I often thought about him. So months went by and I had moved to another place to work and I noticed something white by the side of the rail track and I went over and to my delight was my friend the mouse. He was very old now and his grey fur was white and he could hardly walk, so I saw an old discarded coat by the side of the road. I tore the sleeve out and put my friend inside because he was too old to walk and he had crawled hundreds of yards to see me and he could not walk any further. He was on his own again because his family had gone up now and left him but he had to see me one more time. So I wrapped him up in the sleeve and carried him back to his own little house in the stones. I came the next day with some biscuits and some cheese and placed it in his little house in the stones. I went back after awhile but all that was left was the biscuit packet and the sleeve of the coat and I never saw him again. I very often think of him as a friend but I don't think my friend is there anymore and I will miss him. That ends the story of my friend the mouse.

Blue Scars

I feel I am on a one man mission to illuminate my life as a miner and I know what it feels like to impart this knowledge to a generation that has no idea of how powerful our valleys were to control our destiny, and how it was wrecked by political power that caused the demise of a once powerful industry, but many cannot by the magic of the story teller of the past who try to capture the attention of reclaiming the interest of a younger generation, so that is why I lead up to a story, although imaginative can lead to questions being thrust upon you from the children because a question asked is a thirst for knowledge but looking at the technology surrounding me at this calls room (a far cry from my school days) from the scrubbed top desks and the inkpots sunk into the ink splattered wood by the dipping of overloaded nibs from the pens, from a past era to be replaced now by the shiny tops of the Formica age and the age of laptops that give a greater advantage to a generation of this setting, but I was glad of the basics of my schooling as little as I had after leaving the umbrella of school days, not that you needed much to control the handling of a shovel or a pick. They talk of icons of the ages such as Big Ben, Stonehenge etc, but what better icon can be found than the Davy lamp of the pits? Which was looked upon by thousands of miners as a beacon for their safe return home—that's reality not myths.

Some resent change but have to expect it in this fast changing world of uncertainty, but looking at some of the young faces in my little audience I think to myself 'who am I to judge', they have the chance to hopefully a brighter future than the drab surroundings of a past era' 'best of luck to the kids of the future' I say, but this book is not about what is happening in this present day, but my life as a miner to which I am getting away from it seems! But not before I air my views of over the last eighty years. At the time of writing this book in 2003 in this fast changing world of ours, tales still resurface in social gatherings about pit life that will outlive this changing world of un-certainty and environmental catastrophes that inspires me to write this book as a social document, for a window into the past of which lessons can be learned (thank god—hopefully). Gadgetries that in my day were of the simple but effective way of dealing with life's problems, where we used to worry about tomorrow which doesn't seem to apply to today, because it is too easy to fall into the traps of debts i.e. such things like have it today and pay next year, not realising the implications of the small print. It is easier to read the large print. This is when your worst fears are about to develop, where one's personality differs from others causing the strain and ability of families to cope with their life styles. Anyway I have not changed my way of thinking, a bit too late at my age, because I am one that's not the person to say live today and bugger tomorrow, I may live in the past, but I live within my means.

THE STORY

It was in mid February 1936 and it was bitterly cold where I lived at my house in 3 Vincent Avenue, Nantyglo, but fortunately I lived at the time when coal was king and that welcoming warmth of a roaring fire in which many like myself were loathe to leave at 5.30am into that frost laden air outside and especially when you were not used to the harsh elements that awaited you outside to embark on your first venture into adult life in the pits. After leaving the umbrella of school life I slept in the small box room at the back of the house and there was no central heating in those days, compared with today, although I may say we seemed a lot healthier in them days as much as I can remember. It was a morning I will never forget till my dying day, being awakened at 4.30 in the morning by the shouting up the stairs of my mother 'Come on, or you will be late'. The wind was howling and the windows were covered with frost owing to the condensation of my breathing through the night. I had slept very little during that night and I was tempted to shout back to my mother that I wasn't feeling too well, but then again, I would be making a fool of myself and my father who had approached the manager and talked him in to giving me the job. I must have dropped off to sleep again due to the cosiness of my bed and was awakened again by my mother shouting up the stairs again 'come on or you will be late and you will have the men from the street banging at the door to escort you to work'. My senses were in a bit of turmoil by being awakened from my sound sleep so I shouted 'alright I am coming now'. I was not used to getting out of bed that early so I kicked the bedclothes off me, stumbled out of bed, looked for my trousers and in the process of doing so knocked the clock on top of the dog that was getting under my feet. My dogs was called spot and always slept by the side of my bed because he always followed behind me up the stairs. So then the work started of trying to find my other bloody sock which had gotten tangled up in the bedclothes, I found it and scrambled into the bathroom where the toilet was situated. The oilcloth was freezing under my feet. After doing my toiletries I very nearly slipped on a few drops of pee that had escaped! I got over that and went carefully to the top of the stairs and could smell the nice aroma of the toast that my mother had made. I was careful to tread on the oilcloth of the stairs after mistake in the toilet and I entered our kitchen to be greeted by that red hot fire. I had a wash out the back kitchen and was drinking my cup of tea and was going to have a nice piece of toast when there was a banging on the front door and someone shouted through the letterbox 'come on kid or we will all be bloody late'. I scrabbled into my coat, grabbed my sandwiches which were wrapped up in the News of the World newspaper and was tied with string. I stuffed it into my pocket to keep my hands free in case I slipped on the icy path. The men were half way down the road so I ran to catch them up and I caught them up by Homes Fish shop around the corner from my house. To this day I can smell the frying fat from Saturdays frying—it was still in the air. I caught up with one of my neighbours, a nice chap by the name of Phil Jones (Cider) who had longer legs than mine and I followed him down Garn road, but by Harpers Shop on top of Garn road it was treacherous underfoot because the kids had been sliding on sleighs the previous night, which did not help a lot, so I kept to the gutter. It was a good job that I did not wear hobnailed boots or I would have been in front of Phil Jones. My hands were freezing holding on to the walls as I was crawling down the icy road. Phil looked back to see if I was alright and I often wondered why they called him Cider. Not to this very day do I know but I expect we will all have nicknames or other as time passes. I don't care what they called me as long as I could get to Beynons yard in one piece. My father had already gone an hour before me because his job was that of a hauler, a man in charge of a pit pony and he had to prepare his pony ready for its daily toil, hauling drams of materials through the darkened tunnels of the pit, poor bugger, but at least he was in the warmth anyways. So off we go down Garn Road past Hermon Chapel, now demolished to make way for a doctor's surgery. That is another landmark that will sadly be missed because of its part in local history but we have to move on to make way for future developments. We cannot hold back progress by living in the past I suppose, but to me it was part of my memories but maybe those memories will fade away in time like the smell of burnt fat from Mr. Hones fish shop, but I have to put those thoughts behind me for now and pick my way down to my destination Beynons Colliery and wondered what will await me there. So we came to the Golden Lion public house and proceeded down a more hazardous pitch and more ice than on Garn road. I still kept my feet against the curb stones, I was way behind Phil by now because of his long legs and afterwards I could never remember him catching a bus to and from work, in all winds and weather. I caught up with him when we were on level ground by the Blaina Inn pub, again recently demolished and another landmark disappeared to make way to progress.

Chopper Davies

I remember as a kid I used to go down Trostre by the way of Berea chapel, opposite the Blaina inn, incidentally the Blaina inn had changed its name to a pub called the cog and Petal to signify work and the environment but that did not last long as a pub, but was not any obstacle to progress and as I was saying I used to wonder down Trostre and watch the stream of coal wagons being hauled up to Coal brook siding by two locomotives belching clouds of smoke. These empty wagons would then be shunted into sidings to be taken to Beynons screens, to be loaded with coal but as the last wagon past us we as kids used to jump on the buffers and have a ride up to Coal brook not realising the danger involved but always keeping our eyes open for the shunter because if caught it would be fatal resulting in a smack in your ear or a kick up the backside. These escapades were in 1939 when I was 8 or 10 years of age but that did not deter us until we were finally caught by our village bobby who then reported the incident to our fathers, there ultimately came the belt being unwrapped from my father's waist and that was the final straw. Anyway we made our way down towards Beynons, passing the old baulks of timber, all that was left of old workings of Stones pit which had previously been called the Sum pit and finally came the moment I had been dreading, the gateway into the year and there I was to enter the adult world of the miner and where my life changed forever although I did not know at the time, but I had plenty to think of later one. My main concern at that moment in time was to be introduced to the manager and my knees began to wobble a bit because I did not relish this interview at all but it was a little late now to do anything about it, not that I could anyway, and the absence of my father made it worse because he had already started his shift a half hour earlier to attend to his chores. Fair play, Phil never left me until the time came to be interviewed but that did not pacify me one bit and I looked around at the miners, some were already in their working attire some sat against the walls and a few squatting down having their final drag of their woodbines, perhaps a familiar scene at every colliery waiting to descend to that familiar hole in the ground in which I was to undertake and that started another trend of thought in my mind. I was getting keyed up every second of waiting but looking at the men around me I thought to myself that the men seemed happy enough and seemed not to have a care in the world, so I calmed myself down thinking it cannot be as bad as it looked like, laughing as they were before starting on their trek to the pit head and I wondered what my feelings would be like when taking a similar route because I would be thinking I was leaving all this day light behind me as cold as it was but I had an hour to pull myself together. One thing was for sure I was not going to let my father down and show myself up as a gibber but just then any thoughts were put aside as Phil my friend said it was time to see the manager, the dreaded moment had arrived and by then I had overcome some of my nervousness. I advanced on the door of the office, tapped on the door, and a voice inside told me to enter. I had taken my cap off prior to entering and entered quite meekly. I felt like Oliver Twist asking for his gruel. The manager was standing by the dust grimed window, his back was towards me and I stood there shaking in my boots waiting for what was to come but in those brief couple of minutes that office was etched in my memory like if my eyes wanted to take up all the contents of the office in a single glance like a camera and I wish I could record the interior of that office. I can remember vividly the scarred desk with the old fashioned phone, shiny through use, the dust grimed windows in its iron frame, the dust collected from the years of passing coal delivery lorries, the walls were covered with pictures of pit machinery, their corners curled up by the heat of the office. The calendar on the wall covered with crosses, denoting the dates of the passing days, a pair of old waders by the rungs of the old chair scuffed by the roughness of hobnailed boots, the walking stick that the manager carried on his rounds of inspection down the pit. I could remember most of these items while waiting in my suspense for the manager to turn around and speak. He could sense my discomfort so decided to put me out of my misery which in fact was the worst moment of my life for a youngster as he spoke 'good morning son' I was a bit relieved because he spoke in such a fatherly way realising that it was my first day of employment after leaving school only a few days ago. So in a voice barely recognisable as my own I replied 'yes sir' and I did not say another word after that, so he said 'your father wanted to be present but I told him that I would see you alone and I wanted to know if this is the sort of life you wanted to take up being young and just out of school?' I replied 'that is what I want sir'. By now my voice was getting stronger having gotten over my first fright and was getting more control over my feelings, the manager said 'if you are certain in your own mind you want to try you will wait for the man in charge of you to take you to the lamp station, get your lamp and he will escort you to your work place'. 'You must listen to everything he says because he is responsible for you now, so off you go and good luck son'. So after I had shut the door of the office I put my cap on and was taken to the lamp station and then into the pit head baths to wait for my escort to change into his working attire. There was not any need for me to change because I was already in mine, so we went on our way. I had collected my lamp which weighed like a bloody ton, I expect the weight was 5 or 6 pounds and in addition I was given a damn oil lamp (Davy lamp). What amazed me was my other lamp was more or less the shape of a lighthouse in replica and when I was carrying it I thought I was carrying the lighthouse, but when I saw men with these lamps swinging on their belts it seemed they were totally unaware of the weight, probably they were used to that weight but to me it seemed I was carrying a milestone, and then I thought I won't be carrying it far anyway (hopefully) but I had a surprise waiting for me later on where I had to carry them to my place of work and little did I know that was in the region of two miles or more, oh yes, quite a surprise I can tell you, but I will come to that in the hour that followed after leaving the bath house and the same silly questions eventually arose 'just starting are you boyo' and I thought what a silly bloody question to ask, anyway we entered the baths and I was taken to my clothes locker, on the door of my locker was a tab with my number upon it and the colour of that little desk was green denoting the street when you returned, other colours were on adjourning streets. My clean clothes were in my locker on the clean side to which my father had placed them prior to going to his own job of work. I said to him what if I don't have the job 'well you can always go and pick them up again' he said that was logical I thought but I had already looked in the locker and there was a bag of something stuffed inside. I assumed that was clean clothes but did not have time to examine the contents and if I did I would have found that I was minus a pair of clean shoes, my mother in her wisdom had overlooked this by trying to put together my clean clothes so it resulted in me wearing my working shoes back home after work. God knows in what state they were in at the end of the day's work but it was not a big crisis so I would not have worried about it. I would have gotten over that I supposed, anyway I walked through the passage way dividing the clean side to the dirty side and was comforted by the scene that met my eyes and the thought that went through my head. This was my first entry into the adult world of a miner and what a scene it was. Men in various stages of dressing and undressing by men returning from the night shift previously and what a shambles that was in the smoky and dust laden air of the bath house, bloody hell I thought how am I getting through this lot to get to my own locker but realised there was not any particular rush at that moment in time because I was already attired in my working clothes, so there was not a problem there but I sat at the end of the steel bench that ran the length of the street and watched the pandemonium that was around me. There were a few words of encouragement from some of the workmen I came into contact with saying to me 'don't worry kid it's not the end of the world you know' and 'don't listen to half these buggers around here because they will tell you anything and listen only to your new butty, but you will learn as time passes, so don't worry luv, I was like you once and know what it feels like, but you will soon get over it' and with that he went on his way after giving me a few words of encouragement and a wave of hand. Oh well I thought I wish they were all like him but I was soon to find out that the biggest part of the men had families of their own and respected me for having a go at so young of age. Some of the miners were already changed into their working clothes and were stretched out on the cement floor having a fag in the warm atmosphere of the bath house before entering out in the frosty air outside. Some had even brought their lamps inside to save them rushing to the lamp room last minute. There were young men amongst the more seasoned veterans

Blue Scars

but what impressed me most was the jovial attitude of them waiting to enter the confines of a dangerous world of uncertainty and I was about to join that brotherhood of good relationship that all miners were renowned for to which I was soon to find out.

The different kinds of smells that emanated from the confines of that bath house will stay with me all my life, the sweat and body odours mixed with carbolic soap, not many smells from lux soap I assure you! You could not afford such luxuries in those days, the stale smell of tobacco smoke plus the added smell of men breaking wind, in some cases overpowering to a sensitive nostril of someone just entering from the frosty crispness of the outside air and the pandemonium of men mixing with men arriving from the night shift, trying to find space for themselves to undress while others attempted to dress, the sorting out of mixed up towels and soap dishes. In a limited time they allowed themselves from some who may arrive late owing to buses being late and who knows? Or locals who may have overslept but we were all under the same roof and men were oblivious to my presence, they strolled around in all stages of undress, some with towels wrapped around them, others naked to the world walking around without a care in the world, looking for a mislaid boot that someone had mischievously mixed up with someone else's boots or put into an adjoining street. There were tricks played on unsuspecting chaps rushing perhaps for the cage to descend the pit shaft but was not appreciated by the recipient of that ill timed prank and I have witnessed the short tempers of some of those recipients I can tell you, so I kept well out of the way of these incidents but these pranks were not meant with evil intent but were very annoying all the same and I don't think Agatha Christie would have found the culprit in the assortment of hundreds of miners that used the facilities of the bath house and I am sure the culprit in question would not volunteer his part in some of the pranks that were played!

So I sat on the bench in the warmth listening to the talk that went on around me and words could not express some of my feelings at some of the remarks that were made in my presence. I felt the shock and of the honesty that some of the conversations that were taking place but it was not a woman's world, anything goes I suppose and it was like a private club exclusively for men of the world, well it was in a way because Beynons pit was a sort of family pit where men knew each other personally, it was a good pit and enjoyable to work in such good company but in all walks of life you will find that exception and I can tell you from experience that were a few characters that I came across while working at Beynons and I learned more about life than I thought ever existed! These men made my life more tolerable for working in such difficult conditions and when the men talked about their domestics in no uncertain terms leaving nothing to the imagination, they spoke as if I was not there and it certainly opened my eyes to what was happening in the outside world, a world unknown to me at that time because I lived in a different generation but I think the generation of today could teach them a thing or two, but I think myself not for the best, but that's a long time in my future because I am talking in 1936 and not the present of 2003, in fact what I know of the present I wished that I would have stayed there but I suppose you cannot live in the past but I think lessons can be learned from it, if its only respect for others, but I have not changed my view and still say I loved every moment of pit life and listened to the conversations of our old characters which sadly are not around today to remind of the old social life that I was part of. I know the introduction of technology has made advancement in the world but yet may lead us into a world of uncertainty, even men of today are forced into targets in their work and compete against one another, in my day you got paid for what you earned regardless of targets, if this is progress then I want none of it.

But back to 1936—conversations of some miners did not include some die hards for some were of the strict and prudent men, their life was built around work, chapel and choirs, there is self identification in the personality of quite a few of these chaps. We are not all cap and knee pads, we have the intellectuals who possess that degree of knowledge with that experience who were the vital cog in mining to keep the wheels turning, engineers that are second to none in their trade, we also have the best of surveyors, we have the best of safety management, thanks to our own icon, the small but efficient davy lamp, now upgraded to methometers to name a few of the marvels of mining. Anyway to get on with the story because I get carried away with so many memories it may be boring to some of you but if this version of mining life ever gets published, I assume you will understand, so I will start my life as a pit boy.

While waiting for the man in charge of me I very nearly fell asleep in the warm interior of the bath house and between the cloying smell of adhesive that was used for patches of cloth to patch the rips and tears of your working clothes. This adhesive called No-so glued the patch to whatever was needed in your clothing, plus the hum drum of voices of so many men talking going on all around me, I was glad when my escort came to collect me to escape from the heat and smell. We left the heat of the baths behind us to emerge into a cooler place where you filled your water bottles, we went outside and were met by an icy blast of air from the doorway. I said to my new butty 'I haven't got a bottle for water', 'bloody hell it is only now you are thinking about this, why didn't you bloody say before' he said. 'Hurry up to the canteen and borrow a pop bottle or something'. I rushed to the canteen and did just that and was told not to bother to bring it back, it was a pint bottle from Saxons in Brynmawr. I rushed and filled it full of water and ran after my butty who was half way to the pit top. Finally I caught up with him and we proceeded to the top of the pit. Sandwiches in one pocket and the bottle of water dragging in my other pocket, plus my two lamps and I had not realised at the time that the pint of water was not going to last me for 7 ½ hours in that dust and sweat I may have to endure. Anyway, we had arrived at our destination and stood waiting our turn in the queue and while waiting I stepped nearer the lip of the shaft and I leaned over and watched the cage descending into that gaping hole in the ground until it disappeared into the blackness like a toy match box, to be met half way by the upcoming cage. I could even hear the voices of the men as they echoed in that huge hole. It was then that the butterflies started to flitter in my stomach but I knew that there was no turning back now at this stage. I could not tear my eyes away from that dark abyss that had opened up in front of my eyes, but what frightened me more was I could not see the bottom of that deep hole and wondered how far it went down and would it end like a well and you would be drowned in the water, but that could not be because the men emerging from that iron cage were quite dry so that was one question answered thank god. The men were laughing and joking and jostling one another as they scrambled to emerge. Bloody hell I thought, look how black grimed they all were, like if they had come from a chimney covered in soot, some of their lamps were quite dim, after all those hours burning in the darkness but the little davy lamps still held their tiny flames although encrusted with that sooty covering. That tiny flame had made history in its significance by performing its usefulness in the saving of countless of miners and safety was paramount in the confines of those dark labyrinths of twisting tunnels, so everyone needed our friend the davy lamp. I was awakened from my thoughts by a dig in the ribs from my escort who said 'come on you little bugger, it's no good looking down there, you will be down there soon enough' and looking at the last of the men having their final puff of their fags wishing they had time to smoke a few more to extend our waiting time to enter that dreaded cage, they threw away their fag ends looking longingly at the smouldering stumps with a tinge of disappointment for leaving them, to be forgotten by the time they arrived at pit station to start their 7 ½ hours graft at the coal face. Then it was our turn to enter the cage, there were no risks in doing so and my friend and I were the last to enter the rail lined platform with a grating of hobnailed boots, all 25 of us, because that was the regulated number of men allowed in one cage. It was the last cage before the cages were used for the raising of the coal drams to start the days coaling system. I took a look at the surface men preparing the empty drams to replace the full drams that would be emptied at their delivery point at the screening plant where they would separate the coal from the ash content (muck), then the sudden jolt as our cage was lifted off the kelps holding the cage in position and I had my last glimpse of mother earth and daylight. As the cage started its plummet into the depths my stomach came into my mouth at the sudden drop of that cage. It was the scariest moment of my life, then the last glimpse of daylight disappeared before we were engulfed by the darkened gloom of the pit shaft. I could smell the aroma of the pipe as it was drawn down the shaft by the down rush of air in the wake of the rushing cage, but as we were the last to enter I had the frontal view of the sides of the shaft and by our lamps I could see the wet and greasy orange brickwork that had escaped the dust and oil from the cages over the period of years plus the rigours of changing weathers that swept down and down the open shaft, but even that scene of brickwork structure soon disappeared as we plunged deeper into the darkness. It was then my imagination took control of my feelings, horrible thoughts started to enter my mind such as what if that rope was to break or the winding man had forgotten something that was essential to the running of his machinery, when suddenly there was a flash of light—'what was that' I thought, something must have caught fire from the friction of the guide ropes travelling through those greasy channels, but my mind was put at rest by the remembering of the meeting of the two cages passing one another at the half way stage and that flash of light were the lights of the miners in the ascending cage, then a sudden jolt and we were at the end of our journey. I tried to look upwards by a small gap between the cage top and the side of the shafts wall to try and see if I could see daylight, I was pulled back by my collar by my escort who said 'get back you silly bugger, what if something fell down the shaft, it would kill you, hurry up and get off or you will be trodden underfoot from the boys behind' and with that command ringing in my ears I hurriedly darted from the cage. I stood by a dram of coal and waited for the men to pass and waited for my escort to give me his orders. In that brief moment I created my own vision and as I looked around me I did not have very high expectations of me making a career in mining and I had to face up to the challenges that may lay ahead, but that was a long way into the future and that set a precedence for me to fulfil my obligation to my father in which it was my duty and expected of me to abide by.

I looked around me trying to keep my escort in sight and was comforted by what seemed to me a large Nissan hut with its curved dome and whitewashed walls, but the vision ended as the voice of my escort shouted 'over here kid'. And stumbling through the gaps in some of the drams and avoiding the lockers (spraggs) that held the drams in position before being released to enable the workmen to direct them into their various tracks to the respective cages and to be whisked up the shaft at a speed different to the one that was allowed when men were travelling. Anyway I would not like to be the one to be whisked away at that speed and the smells that I encounter as I stepped off the cage was a mixture of dram oils that were used to grease the axles of the drams for easier running. The smell of horse droppings mixed with urine which emanated from the nearby stables, it was smells that not only smelt but strong enough to practically eat it, that's how strong these mixtures were blending together but they dissipated into the tunnels by the current of air being forced down a downcast shaft, but the lines of drams on their tracks reminded me huge traffic you see on the roads today, you could barely see the diamond patterned shining tracks which directed them to the respective cages which was controlled by the operator that stood at the entrance to the shaft who had to recognise the signals to the surface operator by a code of signals that they understood. Their work started with a bustling of activities of uncoupling of drams and the releasing the flow of that long line of waiting traffic. My escort by then had received his orders and he beckoned to me to follow him into the narrow confines of a walkway to a place in which was a complete contrast to airiness of that curved dome and we entered a different setting of smaller arches of the entrance to a single line rail track which led to the inner sanctions of a working mine, and as I walked in these narrower confines of the tunnel my thoughts were again getting more imaginative not from claustrophobia, to which I was glad that I did not have, but what pressure that these little passageways held to the enormous weight above in which they were subjected to, the thoughts that we were like rabbits in a burrow and I quickly dispelled these thoughts from my mind and it was not the place to suffer from claustrophobia because there was nowhere to run and you were trapped like a rat in its warren in these holes in a labyrinth of twisting tunnels. I was glad to hear my new butties voice to shake me out of those unhealthy thoughts that had entered my head 'come on son let's get going to get out of the way in case the journey ropes start moving'. It was then I realised that illumination and the height of the pit bottom gave way to circumstances and conditions far different from the scene I had just left a few minutes ago from the bustling activities of the pit bottom.

So as we entered further into the tunnels threading our way through the now moving drams which had started because of the delay of me being introduced into the pit bottom register, we carried on to a destination unknown to me at this stage but it made me more aware of my surroundings and created more of my imagination as we plodded on through the velvety blackness that was my introduction to the shadowy recesses where danger was always present. My lifestyle changed automatically during a conversation between the deputy and my butty when the deputy (fireman) said 'make him listen because his is your responsibility now' to which my butty said 'don't worry boss I will make the bugger listen or he will know what!' I had a good mind to tell him he was not talking to a kid of six but I bit my tongue because I knew he had every right to chastise me should I unknowingly step out of line and my life was in his hands so I should be grateful for that. My lamps were getting heavier as we progressed and the weight of my bottle of water kept bumping into my hips. I said to him 'how far is it now' and he replied 'not far' so I had to be satisfied with that, but it was getting frustrating not knowing when our walk would end.

I lived in the past where there was respect for older folk and not the sick society of this third generation. Not all were bad I suppose but to me the world was becoming a cesspit for paedophiles and perverts that roam at random. The dregs of humanity not part of a civilised society that were not known in my day (or were they?) and if they were, they were certainly kept under wrap being a chapel minded community but being a young lad at school, that sort of life never entered our minds. All my thoughts were the ending of my school day and rushing off home because I hated school and the first chance I had to mooch, not realising at the time how important it was to learn, not that I needed a degree, but that it was important like I have already stated, but you needed some of the basic skills to compete with the challenges of the world in general, especially the changing world of today. No one can foresee ahead of us, maybe for the best, or maybe for the worst, but I did not imagine after leaving the umbrella of school days that I would be destined for the pits and the entering of an adult life of a miner. I should have understood that leaving school, one had different avenues to approach but you need those basic skills and facts that would steer me through life's uncertainties, but the old adage was that you learned more out of school than you did init which was true to a point, but you found out that life outside schooldays was that of experience in some cases and a dear form of education not found in books and that is where proverbs conflict when they say look before you leap and the other proverb says he who hesitates is lost. I have always agreed with the former but if you are illiterate to some extent this curtains your knowledge of your basic learning. I may add though that I was not the brightest of pupils, hence I cannot criticise. All I know now is that you did not need any skills when manipulating a pick or a shovel unless you intended to make a career from it, but even so it would take time to adjust to the rigours of what may lay ahead when entering the adult world of the miner and deal with stresses you may encounter. It was a critical time for any school leaver but speaking for myself the pinnacle of my success came later on, as little as it may seem some time later on in my life, but not before the pitfalls I encountered in that time which contributed to my success, but that's another story. I am in my early eighties at the time of working on this book, it may seem boring to some at the outset but my telling of it comes from my heart and not of anyone else's experiences and I can recall most of the events that went on in the valleys where others often fail, because most of the valleys have cause to grieve over the price they have paid to extract those black diamonds and to lose a loved one. I myself have had the misfortune of some of these in meeting them and there are many stories left untold, but the generation of today has never known the sacrifices that were made to bring prosperity to our valleys. That is why my paintings try to immortalise our fathers and grandparents that have faced these problems but in some ways I don't blame these youngsters, they are better off breathing god's fresh air only to be brainwashed by the latest technology that has been thrust upon then and facts are facts, that this fast changing world of ours has gone far beyond our control and no provision for future generations. Today you see youngsters roaming the streets where there are not many places for them to be attracted to, for instance, sporting facilities or places they can be entertained. They are driven to drink and drugs that are available at clubs and pubs, it is so easy to fall into the trap. Yesteryear we had our own ways of entertainment little as they were and laughter seems to have vanished as has the neighbourliness. Today is a rat race where mostly everyone is trying to make a fast buck irrespective of how they get it.

THE PIT

Fairplay, my butty looked after me and I depended on him as the time passes by and he played a big part in my pit life as the years went, but I certainly earned it the hard way. Anyway off we went in that complex of twists and turns in the tunnels, what was to be my home for the next 7 ½ hours. It was a different atmosphere than the scene we had left half an hour ago and left those massive arches at pit bottom, it was low in some places and some of these areas were not unlike the rib cages of some prehistoric animals as you passed by, owing to some structure of arches resembling a rib cage and the white washing of some of the ring sections. We met a crowd of men and horses as we turned into a main transport road and mixed with them, the dust that was thrown up by men and horses made it difficult to see and filled your eyes with dust, the clinking of the shod hooves of the horses echo eerily when they struck one of the iron sleepers in the roadway. You tried to keep your eyes on the ground at least most of the time, and at the same time trying to avoid hitting your head on the overhead sheaves that carried the steel ropes that hauled the journeys of drams that would start working as soon as everyone had cleared from the transport roads, so we went on our trudge into the workings but not before I suffered a few setbacks because on the ground were rollers that carried the trailing ropes of the engines that hauled the journeys, but underneath these rollers were the cavities to let the rollers run free. In some cases they get frequently filled with water from some leaking joints in the pipe line. By watching my head for fearing of banging against the top sheaves I forgot about where my feet were going, hence my one foot fell foul of one of these cavities and slipped into a boot full of water and oil that had collected there and very nearly broke my ankle and I thought what a clever bloody start to the day! After I had released my foot from the cavity which was mostly dry but unfortunately owing to that leaking joint made the hole full of water so much that I had a job of releasing my foot by now had completely filled my boot with the filthy stuff and there was not any time to take it off other than being trampled by men and horses behind me, so I had to stick it on till I don't know when we would reach our destination and I did not get by without hitting my head once or twice either. I felt it more because I had only the school cap to wear and that didn't do much to soften the impact so I had to watch my feet as well as my head, it was a thing that would become automatic as time went on I suppose. I was glad to reach a point nearer our destination and that did not come soon enough, but the men and horses had thinned out now to disperse along different passageways to their various deployments, and a light appeared in the distance, it appeared that we had reached our objective, thank god for that I thought, because in my reckoning we had travelled two or three miles in which we had. As we set down for a five minutes whiff, I felt the various lumps and bumps which had appeared by bumping into those bloody overhead rollers, but my main concern was to remove my one boot, by this time was getting painful and it was getting that I had a job to walk, what a start to my day, it was a painful operation to release that boot because by this time between water and dust my wet sock was cemented into my boot and it took some effort to release my foot where it had clogged up. When I eventually released it someone shouted 'what's that bloody smell, you must be bloody rotten kid'! I admit that the smell was a bit rank from that foul smelling water. When I explained my discomfiture, they laughed at it but I wasn't in the mood for jesting because my foot was raw from the friction on my boot by walking, so one of the men said 'soak your foot in the water' which had collected in the puddles between the sleepers of the rail track from some leaking joints from the pipes, so I did that and although it was filthy the water was lovely and cool for my poor old foot, so I willed the water between my toes and finally washed my sock of much of the cemented dust. I didn't have much time because our five or ten minutes whiff had come to an end and the men started collecting the tools that they needed to commence their work at the coal face. My butty had sorted out his tools including mine which consisted of two picks, two shovels, and a hatchet that was the most important for cutting the timber supports. I had put my wet sock on and fastened my boot by then, my butty said 'take your coat off son and hang your grub up on a bit of wire and tie it on one of the ring arches overhead because the mice may get at it', so I hurried and did what he said 'come one kid for Christ sake or we will be here all bloody day'. So, I grabbed my pick and shovel and with my lamps climbed onto the ledge leading to the coal face. The ledge was about three feet high driven into the rock face of the end of the tunnel to allow clearance for a dram to be passed under the loading head of the belt that carried the coal from the coal face where two men were waiting to receive the first coal of the day, which were hauled away by the little pit ponies. These ponies were replaced later on by a belt that received the coal from the coal face, but at the present I was more concerned with my own predicament of entering the coal face hampered by lamps and pick and shovel but I was shocked by the lowness of the face I had to crawl, owing to the fact the cutting machine had cut the face but left in its wake the piles of small coal Cummings, thrown out by cutting picks in which there were only about two feet of passage ways to travel in. I very nearly balked at this because it had that claustrophobic affect on me, but I was resigned to the fact that I could not turn back because of the other men that was following me up the face, pushing me further, so I had to press ahead to where my butty was working in the middle of the face and I thought to myself why the hell could he not work near the end so that I wouldn't feel so panicky because I could not see him through the dust that was kicked up by the other colliers. I managed to get in hearing distance and he shouted to me to get nearer so he can see me otherwise I won't be able to hear him just now because of the racket the conveyor belt will be making when it starts up. 'Right' I said and placed my oil lamp (Davy lamp) on one of the supporting timbers that were a yard apart holding the top (roof) so I sorted myself out in preparation for starting of the conveyor belt. I could see my butties figure in the dust not far from me but that did not do anything to stop the pounding of my heart in such a confined space and we did not have any face masks in them days and my throat and nose were continuing to get clogged up, but I was sensible at the time as I thought to take my pint bottle of water with me and by not wearing my kneepads which was not an issue at the time, caused my knees to get sore if my knees found a bit of solid ground. I thought well surely they could have found a better bloody job to start with, then all at once the belt started up and ended the conversations that I had between my butty and myself so I was left to start shovelling like hell to get room so I could move more freely and that was not to come for a bit I can tell you because the more coal I shovelled away did not make much of an impression on the mountain of small coal around me. It was never ending as I thought because I could not get rid of the small coal because the colliers in front of me had already filled the belt and there was not any room for me to throw my coal, I was in a bit of a wazzle but I did get enough room to adjust my trousers which were already full of the stuff, I know that I worked hard, but had made no impression at all of getting rid of the coal, God help me when my butty saw the effort that had been made by me, but I could not explain to him that I was doing my best because at one time I got a bit panicky I had lost my lamp underneath the mound of coal and at that moment I was in darkness I must have hit it off the support by my flinging around of my shovel, what a day I thought, because I did not relish the comments that my butty was going to make, but he would have to make allowances. I was getting quite frustrated now and threw my small coal anywhere as long as it left my place, my trousers were getting full of coal again and I was dying for a pee. I managed to get in a position to do that, my poor back was getting knocked about by the overhanging supports, I bet the disks in my back were getting bruised because of the twisting and turning and the screwing about in such a low roof over your head, but I thought I was doing fine. Suddenly the belt started up again, it was starting and stopping by intervals of the changing of the drams at the loading head, anyway the racket of the belt stopped my butty from shouting at me thank god, but I had made enough room now to move more freely and could afford to get in a kneeing position where I could straighten my back, my butty by now had made a considerable impression of his small coal mound and was attacking the main body of the face and he looked over my pile of coal and shouted 'you want to get bloody cracking my lad', and I answered back 'I am doing my best' and I carried on like I was doing before. I couldn't go any quicker if I tried I was sweating conkers and I thought if I am going to have a row well so be it but the time had passed so quickly that it was very near time for grub time and I had done the unforgivable of drinking most of my water and I had another few hours to go yet. I realised what a damn silly thing to do but lacked the experience of judging the contents of your water, that was one of the lessons that confronted me but it was small comfort in my present position, it looked like that I was going to be taught the basics of the coal face and probably more to overcome but its lessons I would have to learn if I was to make any impression to being one of these stalwarts of the pits.

Suddenly everything went quiet and the men shouted up the face 'grub up boys' so men came from their faces of work and made their way to the delivery and by the time I had arrived some of the men had eater their food so I went to collect my own food from my hanging wire and then again I was met with another disappointment and the most heart aching of all—hanging from the wire where I hung my food were the remnants of wrapping paper, in the haste to catch my butty I had unknown to me hung my food parcel under a leaking water point and my food lay in a puddle of water below. I could have sat down and cried as I looked at the mess in front of me, in shreds. My butty must have seen the funny side because he said 'look where you bloody hang it next time, you dull bugger'. 'Here come on son and have some of my bread and cheese sandwiches'. I very nearly told him to shove it where the sun never shines but I was starving hungry so I accepted and also was given some by some of the men who saw my predicament to which I gratefully accepted. My butty said 'you don't deserve it because you haven't shipped enough of coal to fill your bloody cap—you want to pull your socks up my lad and do a bit more than you have this morning, so get cracking do you hear' I didn't answer but thought to myself I have worked as hard as you, perhaps harder, in my frustration and did not know at this early stage what was expected of me, they didn't expect me to do as much as them surely? One of the team said sympathetically to my butt 'I wonder what you was like when you was a kid? so shut up and don't go on to the kid, it is his first bloody day' that gave me the confidence to have another go, but looking at my Saxon bottle I wish it was the end of the shift, but it was no good to tell my butty anything, the mood he was in at that moment. So we went back up the coal face and I found it a lot easier this time because I did not have to contend with a track full of small coal and the air was clearer now, so I really got stuck into it, and fair play I felt a lot more confident and a few complimentary comments from my butty gave me that added zest to pull my finger out I suppose. In fact I was more resolute and determined than before and wondered how my performance would be like by the end of the shift. The belt was running empty now to allow timber to be put on the belt for the making of supports for men who had their coal off and to cover the expanse of ground left open after the extraction of coal. Securing the top was paramount for the safety of the men that would be clearing the debris left by the erection of supports, so my butty instructed me that no way was I to handle the lengths of timber that were being put on the belt but to use the gaps between timbers to clear my coal that was left and I had not made enough progress to warrant any supports anyway. I said something under my breath which was not very complimentary and looked around for my Saxon bottle to have a swig but there was not a lot left in it, my own fault for drinking most of it before grub, so as a consequence my water problem was critical to say the least and sympathy from my butty would not be very comforting I don't suppose considering my disaster at grub time and his attitude. So in the words of the politicians lessons can be learned from the little upsets and in my case you can be very sure it was a very clear form of education one can only learn from experience the only solution was to bring ample supply of water next time that is if there is a next time. I looked longingly at my bottle and I had no excuse but to put up with it I could hear the thud of the hatchets as they bit into the wooden supports and now the air was clearer and most of the dust from that dry small coal had practically disappeared and hearing the activity of the men that were cutting the supports gave me a sigh of relief realising it was coming to the end of the shift, my butty left me alone after telling me to watch the top over my head and should there be any signs of any movement like dust seeping from any cracks in the top over my head I should inform him and go nearer to him because when dust filters through these cracks there could be something to follow Caused by the extraction of the coal beneath it, hence it was imperative to tell him that the part of the roof then could be supported quickly. I thought that was a very sensible thing to teach me, it was that teaching that held me in good stead, later on in life it was one of the best steps in learning the basics of coal face working and made me aware of some of the traps that faces men in such confined spaces. It was then I realised the responsibility that was put on his shoulders by being in charge of me, I respected him for that and it also altered my attitude toward him. He then shouted to me 'you had better come up by me it might be safer and you can watch me making the supports, because the chaps will be up here in a minute to get rid of that coal of yours, it has to be taken care of otherwise the coal cutter will not be able to cut the fact for our next shift tomorrow'. Half an hour went by and all at once there was a rush of picks and shovels working like hell at my place of work to get rid of that block of solid coal. One of the men shouted up to me 'go and have a whiff down at the loading head out of our way kid'. By the tone of his voice he meant it in a kindly way for me not to get hurt, that's the way of the big hearts and camaraderie that prevails in all pits to which I was to learn later on and I was proud to be part of that army of miners, great people that is why this book if ever published by me is to try and immortalise the spirit of what prevails in the pits, but previous to that, my butty had shouted to me to come up by him because he said 'you don't seem to be doing a bloody lot anyway'. I had crawled up by him and marvelled at the progress he had made but being a man (and a strong one at that) with his experience accounted for his advancement into that solid body of coal. I was also mesmerised by how he cut and the manipulation of the lengths of timber in such a confined space by placing these lengths on the floor and by placing the length of the shovel along its length, perhaps a spread out hand to get the desired length that was needed for an upright and the shaving of what they called a lid, to which he placed on top of the upright to form a letter 'T'. The lid protected his back under that canopy of the extended tee section but it was the deft way he manipulated the hatchet which belied belief and I thought to myself if I had performed such an act I would have probably been minus a few toes plus the shaft of the blade under these conditions would have caused myself some injury in the sensitive part of my body! But I suppose all this would come with experience allowing for the fact that I would still be playing an active role at the coalface, not that I intended to make a career out of mining, who knows what the future may hold. This story I am writing may seem boring to some readers but it explains the fundamentals of the facts of what miners have to contend with when men explore the unknown. Every stroke of a pick into that body of coal brings it closer perhaps to another dimension, who knows that lies behind that block of a million years ago because trapped in some of the fossils that have come to light are the imprints of the leaves of perhaps the great forests of million years ago and I myself have proof of it, in the artefacts that have come to light in the process of the extraction of certain areas of the coal bed, but I am straying away by from my original story and I get carried away by the passion I have of explaining in terms that I know of in having spent two thirds of my life in that darkened city beneath our very feet. I may have reiterated parts of this story and the handling of the pen I am writing with doesn't seem so nimble as when I was younger, but I am thankful to be able to write and of illustrating my biography to bringing attention to the men and boys lives in the pits, I am thankful for this chance as little as it is.

So I continue with my memories, I was so engrossed in my attention to what my butty's ability in dealing with his performance that I was brought back to my senses by him saying to me "could you do that son". I replied I suppose I could given time and the experience that you have, but I don't think it would be the standard you have shown me, it pleased him and he mellowed a little after that, he said "let's look at your hands son", I showed him the blisters on my fingers and the scars cut into my arms from bits of fallen debris, now turning blue with the coal dust, and all he said to me was "if you pee on your hands that will harden them it may sting for a bit but it works, my son, you will get over that". So after spitting his bacca out and putting a fresh chew in his mouth he said "you had better get your tools together because the men will be up to get rid of that coal you haven't touched, because we cannot allow that bunch of coal to stop there, because the cutter will not be able to cut the face for tomorrow, so when they come up you get out of the way and make your way down to the roadway where they dump the coal". "Get out of the way son" the men shouted "and get your arse away from here so scoot it". I needed no second shout, I was already on my way down the face to let the men attack that bunch of coal I had not touched to allow the cutting machine to proceed on the night shift. It was then that I noticed the camaraderie of men that worked closely together, I thought it was the finest ting I had witnessed at the beginning of my apprenticeship and I had only known for a few hours, I know I had not made much of an impression, but thrust into a foreign environment of a coal face without any prior knowledge of what was expected of me, but I was determined to conquer and master my fear of that confined space and made a resolute approach to the situation at the first opportunity I had and relied on your own self will to battle that claustrophobic feeling that may develop by the low surroundings that I was to encounter in its infancy, battle I did and eventually won, thank God. Hopefully I would maintain my confidence as long as I stayed in that employment that had been offered me in the early stages of my life and entering an adult way of life.

Blue Scars

I arrived at the coal head to be met by a few of the men that had completed their task and sat down by one of the oldsters that was employed to handle the coal that was received off the belt and listened to the conversation among the men that I had not much contact with during my brief entry into the face, but you can get the subtle humour from some of the characters, the one beckoned to me and said "come and sit by here son for a quarter of an hour and have five minutes, but sit on that sharp stone". I didn't cotton on for some time by his subtle remarks but after awhile I could see the funny side, his remark meant that I wasn't to stop too long. These were some of the characters I would soon be able to meet during my introduction in the pits but some of the lies and fibs that were told were not intended for my youthful ears, but were told anyways. I did not understand them anyway—that was adult talk (but I know now) and as the weeks went by my embarrassment gradually made way for sensible and moral issues and that is why many problems were sorted out and by the end of the day by advice and the support that came out of these conversations by the assortment of the workforce on issues that effected our everyday life I became to understand why there were so much of the friendliness that was relevant to pit life. But my thoughts were only to get my coal off and go home, that's all I was concerned about. I did not have the responsibility of these married men that depended on coal as their livelihood, but to me it was just a job, but I was expected to pull my weight while there and help my butty to fulfil his wage packet and to the delight of him he could see that I was making that extra effort, it also gave me that added incentive to feel that my butty appreciated what I was doing.

But in this fast changing world of ours I suppose some of us have got to change with it. I myself am not trapped in the past but abide by most sensible things that the change has brought about, one of these addictions is the use of drug taking. I myself only take an aspirin a day or the use of antibiotics for different ailments everyone has to contend with but you can get too dependant on those too, but the drugs that are available today by unscrupulous drug dealers are potent and dangerous, drugs like heroin, ecstasy tablets at a price and I have seen the effects of some of these practices subsequently resulting in death, but I am straying off my memories again. I dwell on happier times when coming home from work and being greeted by my wife (Tilly) always ready with a waiting meal and not the sort of dinners that around today, oh no, dinners that are bombarded with atoms from a microwave and frozen foods in the supermarkets that probably have been there months. I am talking about fresh foods that were available in our markets, I can almost taste the pork and beef dripping, the home made pigs head brawn, homemade faggots and peas, but most of what is happening the foods are more conducive and contribute to our cholesterol in our bodies, and the doctors of my day had less to cope with in the way of illnesses. As far as I can remember or perhaps I was too young to notice, but like everyone else we all enjoyed the goodness in those foods that were about in them days when less chemicals and additives were added in our food chain stores, but listen to the scientists to what you eat, don't eat to tomorrow till we find if it's safe or not but carry on (what to believe). Gone are the sing alongs at the old piano stage (to which I was one of them) but I bet the youngsters of today are saying to themselves, thank god for that, but it would be nice to turn the clock back and salvage some of the better aspects and keep certain traditions alive and don't forget some of the forgotten glories still resurface even now, even if it stabilizers the unrecognisable voices that are screamed at you today.

To continue with my story, we were having grub one morning and you could see the scurrying of those little denizens of the deep. Some were grey some were brown and soon the topic of mice was turned around to household pests to which one of the characters of our little assembly had a limp that gave impetus to his story telling to which whether to believe him or not but he told us of his experiences of household pests that caused his limp. He said his house was plagued by black beetles and every time he came downstairs in the night after a couple of pints to have a pee in the early hours of the morning in his stocking feet his feet were always crunching the black beetles that plastered the floor in the darkness of the early mornings. (I wondered how they called Danny Black Pat). So he was telling his neighbour and the neighbour said that he had the same problem till he had a cure for them, so he told old Danny his recipe of how to get rid of them by placing a saucer of beer on the floor so that the black pats would drown themselves in the beer that had attracted them, "right ho" Dan said, "I'll try that to which I did" he said, because he said they were a bloody nuisance. "I could not find a saucer deep enough to I found an old frying pan under the stairs without a handle so I used that because it was deeper than the saucer and would trap more of the bloody things he said. So I put a quantity of beer in it although I didn't like wasting beer but it was worth it to get rid of them and that's what I did. I went off to be as usual and came down in the night to do what I usually do and got to the bottom of the stairs, stepped off the tread of the last step and fell over the bloody dog that was lying at the foot of the stairs and threw me off balance and broke my leg, because the dog had lapped up the beer and was drunk at the bottom of the stairs and that's how I got my limp" he said. His story was broken by someone who shouted "you bloody liar you done that when you fell down the Lion pub steps" but Danny made no comment, but I often wondered was his story true or not, but I was inclined to believe him, you could not prove it anyway. I still don't know to this day because he was so serious but his story was being capped by another character who claimed his butty was so drunk after drinking in the Black Lion in Brynmawr that he fell against a lamppost in the middle of the square, he held himself up and started to climb the lamppost and shouting up at it "I know you are in because I can see the light on upstairs" he was finally arrested for disturbing the peace, but he slept all night in the police station, he said a few other exploits but we did not have time to hear them so that ended grub time when my butty shouted "oi kid what about making a start that lot will keep you there all day if you let them" and I believed him too, but you had to tear yourself away from that lot to get down to the serious matters of the day and that was that although I was loath to leave the humour of those characters but it shows you the friendliness that miners possess that makes work places more agreeable to work in, but you also have to have a quality of authority too, to stabilize working procedures such as measurements of one's work, the controlling of air flow etc. it is not all pick and shovel, but anyway let's get back to the basics of working practices, so after I left that assortment of characters and I am quite pleased that in that twenty minutes that I listened to fellow miners in their different variations of their stories, made work more tolerable

Blue Scars

so one day while having food the man in charge of the coal face the man that was taking the responsibility of assessing work being accomplished by individual miners at their allotted tasks, this man was called a puffler (head man) that reports his findings to the deputy (fireman) who relayed his reports to the management time keepers who assess work done and pay accordingly of work done at the coalface, this puffler came up to me one day and said "I don't think that you are cut out for this sort of work Johnnie are you" I replied that it was only a few days that I had time to get myself sorted out and to adapt myself to the working of the face and had not much time to prove that I was capable of doing the job I was expected to do, but it was the way in which the question was asked that needled me and the implication behind his questions and his not very welcome smug smile, to which I took an instant dislike to and I thought I was not going to get very far with that type of attitude and that he was only one of the men that had the same job as the rest of us, but apparently took the job of responsibility on his own back by tittle tattling to the officials and in one term I can describe as a bosses man, he was well known for this to which he was not in favour with the rest of the crew, including myself but given the type of man that he was a little bit of power and it overrules his once good intentions and also his personality, and it had acquired him the nickname of Captain Blood in his attempt to get some form of status but you could see the contempt the men there had acquired, but that's by the way, but his approach to me is what rattled me, but it put me on the alert but it worries one to what was his next intentions would be, and it put more pressure on my efforts to pull my socks up although I don't know what else I could do to increase my efforts because I had pulled all the stops out that I could manage, ok, bugger him I'll carry on like I was always doing it's up to him if he wants to question it, that's what I felt like, but you would think that he would give encouragement not dishearten one, but if your face doesn't fit to his requirements then you would have a hell of a job to dissuade him from his stance, but he did upset me there was no other way to put it, so I just carried on, I would find out in time I expect but he thought of nothing but coal, coal, coal, coal and if I thought in fairness to my butty that if I was a liability to him and the face production then I should go and I said so to my butty and told him of the conversation that Captain Blood had with me "take no bloody heed of him but if you think you cannot cope "then it will be your decision to make and not his, carry on kid" and that's what I did for quite awhile, but the Captain Blood incident was always at the back of my mind and the thought of being an incumbent to my butty who had a family to keep played a little on my conscience and in my own mind I was not robust enough to enable me to continue such a task at the face, so I was in two minds to say to my butty, own up and tell me am I a liability to you, but he was a gentleman and he would not put himself in the position to say that I tried hard enough but did not live up to his expectations, and was fair to me to say so, to which I did when the time came but was always afraid of taunts from the older boys "too hard for you was it butty?" but it was a chance that I had to take plus the smirk of Captain Blood saying to himself I told you so but that's the way the cookie crumbles, but at least I had the satisfaction of saying that I tried and that held me in good stead amongst my friends at the face, so I told my butty my decision and his reply was "it's up to you son, if you think it's what you want, it is better than killing yourself on a job that you don't fit into" but I had not learned the lesson of controlling the water problem of keeping some back for finishing time. So one day I hung back behind the men going home at finishing time, thirsty, I opened the valve of the water pipes that was overhead, the water tasted brackish and of rust, which was expected by the sediment left in the valve, but it quenched my thirst. I didn't notice that one of the men behind until he spoke "don't you ever do that again son, if were that thirsty and run out of water, ask one of us, do you hear" and why don't you bring a bigger bottle" I replied "you are right" it would be better I thought I've got enough weight to carry now with these two lamps without having to carry more but it was the only solution, but it was alright for him he was a big chap but after finding an old haversack in the rubbish bin decided to carry a flagon bottle. The reason before was I could not put a flagon in the pocket of my school blazer but I often wondered later on how did the smaller kids of 10 and 12 managed when they had to crawl through narrower passages than what I worked in, because that was the ages of them poor little buggers in the past, dragging wooden boxes through passages only two feet in height, these boxes were called curling boxes. Women also worked in an undignified way by being door women opening doors to let the little ponies pass through with their burden of drams of coal after being filled by these curling boxes, dragged by these young boys and I thought to myself what I thought was work was nothing compared to what those little pioneers went through, but early in the 1900's the ages of boys rose to 13 and 14 and I realised I was only 14 myself.

Anyway we started our way out to pit bottom and home through the dust that the ponies kicked up in front of us but I had learned my lesson by now remembering my boot full of water and my bruised head, that was my introduction to these little pitfalls that I had encountered during my entry into the mining world, so I started dodging those overhead rollers and the sheaves between the rails that carried the haulage ropes, the overhead rollers were my main concern because by school boy cap was no protection from them, and I knew I would be black and blue from the face and my face I expect was as black as the ace of spades from being in contact with all that dust and I thought to myself if this is mining I want no part of it.

The background of our industry does not make very good reading, it has a long history of toil, sweat and tears, occasionally death but there is no substitute for experience, it was a clear form of education in mining and other industries that you work in, but mining is a dark underworld city where you don't have the advantage of God's sunlight, so this is what I had been designed to do and not through choice but necessity because there were only the steel works and a few factories but not for kids like myself leaving school, but when you become a miner you become part of a family, a family you had to respect and put trust in as I found out and it will live with you forever and you become one of them and from the time I worked with them I made many friends and the families I came in contact with, so I walked and walked and I thought we were near pit bottom when I saw this light ahead of us but it turned out it was a light coming from one of the many little haulage houses shining on the rails ahead of us, they are was cleaner now, owing to the velocity of air and it was quite fresher now and much more stronger as we neared the pit bottom and I could even see the shape of the ponies in front of us now so I thought we were not far away now, bloody hell I didn't think that we were that far inside, there was a turning ahead of us and as we entered the turn there was the welcoming sign of pit bottom, and what a relief when I joined the queue of men waiting to ascent that shaft.

My only claim to fame over a long period of time is that my work has been shown and televised by different channels of ITV and BBC including Nicola Heywood Thomas series of High Performances, BBC series of Weird Wales and of having the pleasure of the company of Rebecca for the period of approx six hours of filming the Rebecca John. I am referring to a Newscaster for ITV and also one of my paintings was introduced by Sara Edwards the newscaster for BBC Wales and narrated by Geraint Vincent news reporter who suggested that one particular painting had particular residence at my home, it was a of a mystery painting that has baffled the media and national newspapers alike, to find a solution it contained the image of miners trapped in the painting that defies any explanation, it also baffled Roy Noble in my interview with him at the BBC talk show on the radio at Cardiff about the phenomenon of that painting and its ghostly appearance of dead miners and still defies an explanation. It still attracts quite a few people to my home, this particular painting has dominated my work, of which I collated approx 400 over the years I have spent as a mining artist not counting the countless hours of approximately four or five hours a day over thirty five years which is no mean feat of determination to foster my desire to continue my art work, but time I am afraid is running out now because of age, but as long as I have the will to persevere and capture the imagination of the generations to come, I would be satisfied.

Chopper Davies

But before my entry into the art work and after leaving school I found it boring because you miss the company of the other kids and could not find anything to occupy the restless roaming around ash tips and occasionally accompanying the local greengrocer's son on his regular rounds of his horse and cart. His father kept a small confectionary shop on Garn road by the name of Mr. Prosser, a little shed of about twelve by eight if I can remember but it served its purpose to our small community, he even served paraffin and oil in a little stone shed alongside the shop. Mr Prosser was a fine gentleman to us kids and was very popular in the community. I can still smell the mixtures of sweets, mixed with the few vegetables and a whiff of the paraffin as it wafted through the door when someone wanted the oil for the lamps, it was an agreeable smell and will remain with me to this day, and the Sunday morning jaunt for the small bottles of football stout pop to quench the thirst after your Sunday roasts, more appreciated than in the minerals ones that are about in supermarkets and public houses, a far cry from the nettle pop and homemade wines, like elderberry and other of natures remedies that cured a lot of the ills that affects us today. I was only 8 or 10 at that time but my memory is still fresh for those unforgettable days as a kid, but at the age of 14 I found another way of occupying my time, by travelling around with Mr. Prosser's son Elvert, who ran a horse and cart service of fruit and veg to outlaying farmers and neighbouring houses that were off the beaten track. Quite a thriving little business to which I used to deliver to houses that had their regular orders any my only pay from these errands were a few cut apples and quite a few juicy plums and strawberries that I used to pinch when I followed him home in the darkness, hence I believe he knew that I was pinching them, hence the absences of any pennies that I might have received but I thought you cannot have it both ways I suppose. Anyway I stuck that for a few weeks and got fed up of getting soaking wet most of the time, but my big break came when my brother and his cousin were working in the city of Bath, they were working as brick labourers and asked if they could, would I like to go, to which I agreed, but it was up to my mother and father to agree to it. Eventually one weekend they came home and said that they found something suitable for me and would I go, I said I would but it was after my parents agreeing to it and so the date was set for me to accompany them to Bath. I was excited because my mother was glad to see me go if only to get from under her feet I suppose and secondly it would stop me from scrounging around the ash tips, so I decided to go back with them the next time they came home, they did not say what my job was to be, but on arriving at Bath I was told I would be a hairdressers assistant, in fact it boiled down that my job was to be barber boy of which I had no idea what that meant but soon found out. Later on in my apprenticeship but before I undertook the post I was to find lodgings but was agreed that I would stop at their lodgings and was introduced to their landlady of the house in which they discussed lodging money, agreement was reached to what earnings I would receive at my job. So after settling down I got acquainted with the family who to my surprise was the local verger of the church adjacent to our house, it was a very imposing looking frontage to the house known to the affluent society of the town as a very respectable part of their community and I wondered how I was going to adapt myself to living in a vicar's house, but was soon adapted as one of the family, being so young, but fortunately they had a son near enough to my age that did not conform to his parents expectations sadly enough, but to me he was my saviour because I could speak to him on level terms against his mother's wish to show an example to a stranger, but I think those motherly words to him, fell on deaf ears because that one had a mind of his own and would not be converted to the ways of the Lord it seemed, but it was ok by me, but kept myself at a certain distance not to give my brother any problems, I had enough sense for that thank god, anyway my lodgings rental was 10 /—a week, 50p today of course, but I was only earning 9 shillings a week so my landlady allowed me one and sixpence back for me to buy comics, or go to the pictures in town, but then it was only one penny for the tram ride to town to which I used for the time I was there (I wasn't there that long because I became homesick with not having friends). Anyway I was there as a barber's boy. On Sunday mornings I used to watch the parishioners entering the church through a bay window of my bedroom, which overlooked the church yard and observe the people in their fine clothes, in their starched high collars and the women decked out in their finery with their parasols and the kids in their bonnets. I thought bloody hell, I don't think I could stick that and I'll bet the chaps were glad to go home to take them collars off. Like I said I lived in an affluent part of Coronation Avenue of Old Field Park, Bath.

I prepared myself to get my act together for being introduced to my new job at the hair dressing salon and I was a bit apprehensive about it but was put at ease by the landlady taking me, I was employed by Mr. Redman of Coronation avenue and I was shown the customs of the salon which was not too complicated thank god and I had to get used to different accents of some, my one particular customer was the local bobby who when I visited the town centre of Bath I was told by him to wait in his little sort of sentry box that was placed in the centre of the main thoroughfare of the town. He would then see that I arrived home safely because he himself travelled on the same tram to his home nearby. One day I went to town, I was fascinated by a group of people that was descending the steps of the Hotel Royal in the centre of town. They were attired in funny clothes to which I now know to be robes, that was my one claim to fame at the time because the central figure I learned from my new friend the bobby was that it was Haile Selassie of Abyssinia. This was in 1936 and I have kept that in my memory every since and I can see my friend now with his white gloves on directing the busy traffic. I did not have any supper when I used to go back to my lodgings because I used to share the sandwiches and a cup of tea and biscuits with my friend. So one day I will share with you at a later time my exploits at the hairdressers and my adventure as a barber's boy.

Into God's given light and air the men were sitting on tops of drams and some were in various poses waiting to be rose to the surface when the time arrived for them to do so after the pit bottom men had finished their jobs of stabilising the movements of drams. Because pit bottom workers were the first to start and last to finish so I have nothing to grumble about compared to those men handling the rusty shackles (couplings) for seven and a half hours. I bet their arms were ready to drop by the end of the shift but after years of working at their job, they had adapted to their kind of work and the exposure of those biting down rushes of cold air from the surface. I looked around for my butty but there was no sign of him, because he knew that I would be in the company of the colliers of the face, the movement of the queue was at a snail's pace because you had to queue of about a hundred men to be counted at twenty five to a cage because that was the number of men allowed on one cage, but it seemed to me as if I was waiting in a queue at the local cinema, that's what it looked like anyway, but it started moving a little faster now and the afternoon shift colliers were starting to come into view to start their toil at whatever jobs they were directed to, and you usually get some jocular remarks as they passed by you. Such remarks like "stop down here boys if I were you because it's bloody snowing on top". I thought to myself I don't care what the weather is like let's get out from here and some of the crafty ones making their usual voices heard by saying "anybody got any bacca to spare because it will only go dry in the bath house" or one saying "damn it I must have left my bacca in my other clothes" but most of it fell on deaf ears because it was the usual form of greetings or shall I say craftiness of getting a chew of bacca. Some of these use to cadge enough that they did not have to buy their bacca the next day, but it was the usual day in the pits with their humour, so I waited for my turn to arrive to ascend the shaft but most of the men in front of me were waiting in anticipation to have a drag of a fag they had hidden away at some secret little place, hoping that some other chap had not found it. Cigarettes were not allowed in the deep mines such as ours, only chewing tobacco in sticks which looked like pieces of rope wrapped up in silver foil and I thought how the devil could men chew that filthy stuff I don't know. Loose tobacco may be allowed that was acceptable but in combination to the means of lighting it, it may give cause for concern and result in dire consequences which could happen in deep mines so the searching of contraband at pit heads would deter such thoughts so it was imperative that those sort of thoughts must be dispelled for the safety of the miners in such a dangerous environment, but there was talk that in some collieries of men concealing cigarettes in a plastic bag among their food in their food tins or in water bottles but in the forty odd year I have worked down the pit at Six Bells I have never had that experience of the men at my pit thank god, but the only thing that I find, but had shall I say, made a law to screen people as to their personality but relied on their own common sense to conduct themselves to what the perils that are involved by doing acts that contravene the mining acts, but unfortunately these risks are and will always be with us. But anyway it was my turn to enter the cage after passing the man who checked the amount of men that was allowed on my cage, I was the twenty fourth or fifth I think because I was near the gate of the cage, to which you had to turn around to face the gate to allow you to step off in that direction, and I was lucky to be in that position near the gate and not get squashed in the middle. Looking out I could see still a number of men waiting their turn to follow us up the shaft, then the sensation of the sudden lifting of the cage and we were off to the welcoming sight of mother earth daylight. Darkness then as the cage gathered speed to the centre of the shaft, a flash of light passed us midway by the descending cage of men of the afternoon shift and as our cage neared the surface, daylight revealed, the sides of the shaft, greasy and wet with a few bits of orange brickwork, that had escaped the rigours of years of weather that had been exposed by a open downcast pit, I was saying under my breath "come on, come on, hurry up for Christ sake let's get off in case something happens" like the rope breaking or something goes wrong with the winding engine. Thinking of that gaping black hole beneath us I suppose I must be echoing the thoughts of many early starters that entered the pit for the first time, but at that moment in time I was only thinking of myself. At last I could hear the murmuring of voices at pit stop and the aroma of tobacco from the bandsmen's pipe and smouldering dog ends of fags the afternoon men had thrown away after their last drag of their fags, we were met by an icy wind as we cleared the tip of the shaft but I did not care what the weather was like as long as I could get off the cage. I was nearly bowled over by the rush of bodies that was behind me but I needed no pushing, I was glad to be first off, I followed the men to the baths but at a slower pace because I did not want to get mixed up with that rumpus with men scrabbling to get to the showers first, trying to get the first best shower because some of the showers were not all that good, so I strolled over after I handed my lamp in, I hung about for 10 minutes for the crowd to disperse and entered the steaminess of the bath house and made my way to my locker street, remembering my colour was green and I had to think what my locker number was but eventually remembered it. I thought to myself I will have to put a chalk mark on the door to make it easier for me to find and that's what I did, I sat on the end of the steel bench waiting for the main body of men to vacate the street and looking around at these unfamiliar parts of this large area of lockers and cubicles for showers and wondered how many men had passed through this building over the years and how many of them had failed to return. I thought it was nice to be alive to value life itself. I tried not to be morbid in my thoughts but I did imagine the men that may be working at my place of work I had just left but on the afternoon half the men would be maintenance men and repairers of some of the main roads and airways that kept that flow of air circulating around those labyrinth of tunnels. Mechanics maintaining the machinery that was essential to enable the machinery to continue the work that was required of them, men shifting conveyors into a new track after coal extraction, pipe fitters

Blue Scars

doing the same work to allow water and compressed air for coal cutters to continue their work of cutting the coal face, they call these men flitters (or turnover gangs) it was heavy and demanding work in confined spaces of a coal face, but every man to his own job I suppose, little did I know that I would be one of them later on, in life, and would be one of the tasks that would be required of me, but I did not, and could not envisage the pit falls that I encountered by that time.

On entering the baths I had emptied the dirty water from my bottle and had a good drink of clean water which tastes like champagne after the rusty water I had drank, but it made me realise how precious water really was, it was a good survival lesson that I had learned in the dust laden air of the coal face, it was one of the hundred lessons that had to be learned in my stay at the pits. I must have hung around long enough for the bath house to be cleared and quite a few of the men had dispersed, there were a few later comers so I waited for them to go, so I had a whiff on the steel bench of lockers and enjoyed the heat of the bath house and I left quite cosy, so much that I must have dropped off to sleep and was awakened by the wishing of the hose in the neat street by the bath attendant and I was still in my working clothes, there was water seeping underneath the lockers by the force of water, it was a good job I still had my working boots on or less I would have soaking wet feet again, then I could hear the attendant shouting "you had better hurry up you little bugger or I'll turn this bloody hose on your, I want to get on with my work", I replied, "I have to go to the toilet first Sir", "well hurry up then and don't belong". I did my toiletries and undressed in a hurry, hung my towel on the locker door and by then my feet were soaking wet from paddling in the pools of water under my feet, then I realised I had no soap and flannel, oh hell I thought, my mother had not put a flannel and soap in my clean clothes, well I could not blame her for that, she could not think of everything, and especially an item like soap and flannel, but it put me in a bit of a wazzle of how I was going to bath, the attendant had stood waiting to walk our street and said "are you ready son", I said "I can't bath because I haven't any soap and flannel", he jokingly said "oh well you will have to go home like that then wont you", but he could see by my face that I was worrying and said "I'll see if I can find something in the attendant room and if anyone has one behind you can borrow them, I don't suppose you want me to wash your back as well would you, because you will be bloody asking that next". I didn't answer, I better leave things as they were, not to antagonise him any further. Having found what I wanted he went and did some chores he had to do in the boiler house so I grabbed the chance of going to the toilet again, and when I returned the soap dish was not there, now I was in stimmer, like if everything was going against me, but it was my own damn fault, so I went to hunt for the attendant to explain what had happened, I hunted everywhere but could not find him and returned to my street and lo and behold the soap dish was back with a wringing wet flannel, so to my delight, I continued bathing but whoever had borrowed it could not have washed very clean in the time I tried to find the attendant, but I appreciated its return and quickly ran into the showers, the men had all gone now, thankfully, so it gave that bit of privacy and I would not feel embarrassed bathing in front of grown men, thankfully my father had gone home an hour ago so I was more at ease, I could hear a few stragglers in the next street singing and talking so I thought I am not the only one

Blue Scars

holding up the work of the attendant, who I could hear swishing his hose around these streets, and I hurriedly washed to get out of his way, I didn't fancy the cold water of that hose reaching my naked body. I wasn't all that black with dust because I had started with clean clothes and they had not had chance to be completely saturated with the filth of the pit, it was only my face and my feet that had really come in contact with coal dust, but anyway that damned hose was getting closer and I got out from the showers a bit quick into the now deserted clean side, where my clothes were lovely and warm owing to the concentrated heat of the bath house, but I had to be careful of the puddles of water left by the washing of the floors and you could not avoid stepping in some of them and it was pretty dangerous when placing your feet on the smoothness of the steel bench, so I took my towel off and placed in the area I was going to dress, but it was my own fault I suppose because I should have bathed earlier, avoiding the discomfort, but we learn as you we go along. So after dressing I handed the soap and flannel to the attendant's room, made my way to the exposure of a brisk and icy chill of the outside air, finding more so after leaving the confines of the warm bath house and my hair was not completely dry in a hurry to get from that hosepipe, and started to freeze rigid after walking the two miles home. My father and the rest of them were already home and sitting by a roaring fire by now, I knew for certain that one man was home that was Phil Jones (Cider) with his long legs, I passed very few people on my way home, I expect they had sense to stay indoors by their fires and went out only to get their essentials and the only little vehicle to pass me was the little red van of Brooke bond tea, with its solid tyres and today would look like a little dinky toy, but was essential to replenish the village shops and was o his way I expect for his last rounds, but at least the driver was in a warm cab. Eventually I arrived home, my feet were freezing and when I had taken my shoes off I very nearly shoved my feet through the bars of the fire. My mother had prepared my dinner, my father had eaten his and was having a puff at the favourite old Woodbine, he enquired how I felt after my day, after satisfying him with the details I had my food and went and had a lay on the bed for half an hour, too buggered to go out and it was warmer in the house, and I had a few comics and Western magazines to read, so I didn't bother to go outdoors, but I came down later on, had a cup of tea and could see the sticks were cut for the fire in the morning in an old biscuit tin. Memories like this will always be remembered but are replaced now in these days by the flick of a switch, the gas age, but will never replace the good old coal fire. I could see that black monster of our black leaded grate in which I helped my mother to black lead it, with zebra polish till you saw your face in it, all is gone now, but it will not erase it from my memory of those homely times at home, so after a few cups of tea and a bit of toast I went back up to bed to read my cowboy books, till I fell asleep to be awakened by the routine calling of my mother's voice "come on you will be later", so a trip to the bathroom to have a pee, downstairs for tea and a piece of toast and to wait for Phil to call, to have his company to a walk to work, but I had a surprise this morning, instead of my food wrapped up in last night's Argus there in front of me was a bright new oxo box with a rubber band around it to keep the lid on, and a penny to have a piece of cake from the canteen put inside as well if there was room for it to be squashed into. So off we went after the men shouted by our gate "come on kid" but my feelings were a bit lighter now, knowing that I did not have the fear of meeting the manager that was one good thing, I arrived as usual and left my food tin in the canteen to be picked up when I rose my lamps and as I was leaving the canteen I heard one of the ladies saying in conversation with the dinner ladies, who said poor little bugger, he don't look strong enough to carry those heavy old lamps and I was thinking how right she was, but I can remember some of those good women, one whose names were Mrs. Best from the top end by Blaina, and another by the name of Mrs. Madden, from Cwmcelyn. Mrs. Best also had a son, a bigger boy than I but was a little older perhaps and worked in a different part of the pit and was employed as an engine driver that hauled the coal drams and was not anywhere near the coal face. Mrs Best was my best friend who was stout and jovial. So off I go on my second shift but I was not so stressed now, but still dreaded that cage but as time goes by it will seem like getting on a bus I suppose, and I had adapted most of the skills of the coal face and had overcome my nervousness and a way to control my water supply, and when given a job by your butty and you succeed in doing it to my butties satisfaction it gives confidence and motivation to do what was required of you in your work, plus quite a few safety values involved. I learned a lot and it held me in good stead in later years because it is never too late to learn. A question asked is a step in the direction of more knowledge and I was captivated by conversations of the older elements, only limited by the coal mines and stories of some incidents through the past years, tales of some unsung heroes that go un-noticed to the outside world, but it was all part and parcel of the camaraderie of pit life, of which I would be taking part as time wore on. But after leaving school it was like trying to break out of a shell of school days and into an adult world and could be in worse conditions I suppose, but it had to start somewhere, why not here, but like I have stated I don't think a career in mines would suit me, but there were no other alternatives, only steel works, which needed more than the knowledge of pick and shovel, but every man and women to their trade and their destiny, at least there were no bosses to keep looking over your shoulder, where I worked you had a job to do and was paid by results, but my thoughts as a kid at fourteen could not settle to view any job. I found that out when I acquired that job of hairdressing at Bath and the 10 bob as barber boy, breathing in the foul breath of different customers, an unhealthy job that was part of your apprenticeship, it was not for me. At least when you entered the pits there wasn't any need for special qualifications just dig, dig, dig, the only needs in them days were the value of your money, where a penny was a penny and the food more nutritious and money was spent on decent food because housewives made their own bread, you could smell the newly baked bread everywhere, it was more wholesome than the added additives today, bread that you are forced to eat today, them days people had their own tins with indentation made perhaps with a nail, topped with a hammer into the sides of the tin denoting what household they came from. When I went to collect our tins I could not resist the brown crusts on the top of the bread and I could hear some of the boys mixing the tins up and some shouting "oi that's our old gels tin" because if a tin had only one syllable on with a single letter it could cause confusion, you can imagine a tin with a 'J' it could mean Jones, James, Jenkins and so forth, but they soon altered that I suppose, but the crust was always the attraction, and in my case it was half eaten by the time I arrived with our tins to the bawling out of my mother and a clip across the ear, but bread and cheese were the staple food of most miners, and my favourite was a sandwich of cold chips or

Blue Scars

scratchings from the fish shop or beef or pork dripping left over from the Sunday meal, it was food that would sustain a man for seven and a half hours, but that is all gone now since the demise of the pits. I believe the causes of half the illnesses today is food from fast food supermarkets with their synthesis of frozen foods perhaps laying in the fridges past their sell by date and the public have no option to buy. God knows what's coming next, perhaps the scientists will come up with something like tablets to substitute pork chops or something I expect.

Back to reality now and my week's work is coming to it's end but I had the feeling that my days at the face was about to come to an end, because of the pressure put upon me by the attitude of infamous Captain Blood and was not conducive in keeping me to cope with the stress of the coal face, but all my doubts were on the way out and I knew deep down what to do and say to level my confidence building and to help identify what kind of work was suited for me, but there was not much of a selection to inspire me as an inspiration to make a career of the pits, but can concentrate on matters that are present at the time and patience not pressure could be a winning tactic against the badgering of this puffler who had been giving me a hard time, giving the man a sense of power, although not much alters a man's personality and he seems to be drunk with power that he had been given. The men on the face had given him certain names, that only a birth certificate could prove otherwise but I avoided any confrontation and tried not to rebel against decisions that were made by him, but there was only one thought in that tiny mind and that was coal, coal, coal. Anyway my feelings at that moment in time was that my time was limited at the face and my face did not let onto the puffler way of thinking, so I knew I was ready for the chop and it wasn't long in coming about when asked by the deputy (fireman) in a kindly sort of way and not the attitude of the puffler, "if I still feel suited for the job that I was doing or would I prefer a lighter job, not that the men on the face had complained, especially your butty, but think about it kid, it will not be held against you if you decide. I have got boys of my own and I am thinking about the best interest for you for a young lad, so think about it son". He didn't put it in an official kind of way and I thought a lot about the way he explained the different jobs of a colliery that would be best fitted for me and away from the dust that I was putting up with and It made me think of a way of saving face without showing that I was a quitter, because as young as I was I had a sense of pride knowing that my father had found a job for me, but he would probably understand anyway, I made a bold bid and approached my butty as to whether I was holding him back with his work "don't worry kid about it if you get a chance of bettering yourself then go for it, that's all I can say" and he also said to me "I know I have played hell with you on occasion but it was for your own good, but you have got a bit of principle to apologise to me, so if you think and decide you want to then so be it and look after yourself do you hear". I told him I would make my decision next week but I knew only a week had passed by of how I would miss all the bantering and jokes played on me and he started to laugh "and in a way I am sorry to see you go but you will be out of this bloody dust and you will live that much longer" but I knew I would miss some of the characters because they have said things a lot close to comfort oblivious to the fact that I was just a kid, but most of the conversations would range from domestic issues or perhaps from their night in the pubs previously, tales about their travels which you could take seriously but it broke the monotony of the day when you had your grub times but the younger element brag about their conquests of the week, but what often amazes me is they mostly talk about women when they are down the pit, it mostly crops up because it's a man's world down there and perhaps afraid to say things like they do in front of their wives or their sisters, but after work and in the pubs they start talking about work and some of the older miners get pulled on by the leg pulling of the youngsters. I have seen myself, grown men being goaded on after a few pints on their sides by the side of the benches on their elbows with an imaginary pick to show the youngsters how low they have to work under after being egged on by the regulars of the pub, of course the recipient of the leg pull had enough beer to last the night perhaps the next night by telling the landlord, leave it over the bar till I come in tomorrow night (better beer then) and some of the lies they told their wives if they were short in their pay packets, some of them would say they had to pay for buckets of compressed air, or paying for rails to be bent, but not all wives were as gullible, they had an idea where the money went, a bet for horses perhaps or a few pints at their local pub which had to be paid by the end of the week and the only way wives could retaliate in a way wives know how, but there were quite a few church going people among the many miners that worked in the pit who would frown on some of the sayings that were said by non going chapel people who spent their money the same way as they earned it, the hard way, some of these God fearing people would turn nasty at some of the crude jokes that was always cropping up at grub times, but as it was a man's world some were oblivious to the beliefs of these religious work mates but they tried to keep some of the conversations out of earshot, in respect, and try not to be in the vicinity of these chapel men, but they could not avoid it in some cases so they had to put up with it, and put up with the disapproval that was glanced at them, myself, I liked to listen to some of these conversations to broaden my outlook in the outside world, but did not understand half of what they said (but I do now). I was often afraid of listening to this pit language because I may slip it out at home and that would certainly be an embarrassment and would be a long time before I would be able to live it down, but it taught me to realise to pick my words at home and I wouldn't think of such language to be used in front of my mother and father, but it was hard after being in the company of men for seven and a half hours to eradicate some of the words used by them.

So Monday morning came around again, the same old routine as always, the walk to the pit with Phil and the men, the changing of clothes, the usual smells of carbolic soap, the bantering of the work force and the buffeting of the men around me, but I was not waiting for my butty now, because I knew my way, picking up my lamps and collecting my oxo tin from the canteen, the filling of my flagon with water, the old haversack I had picked up from the bin was a life saver to me, all these little small chores came automatically to me now but I made sure that I was down the pit before the final hooter went to begin the shifts coaling for the day. The feelings and the dread of that cage was slowly fading away but the thought was still there but it was automatic now, but on arriving at pit bottom the usual procedure was, broken by the sight of my butty, standing by the officials cabin, so I thought, oh oh what is happening now, because my butty would have been on his way in to work by now, but he shouted to someone inside to say I had arrived, and he called me up to the cabin door and said half a whiff by here so I stood by the side of him, until the deputy came out and approached us. My butty said "it's okay kid you haven't got the sack" and smiled at me, I thought to myself I knew I hadn't the sack because my lamps would have been withheld, but there was something going on, that was one thought I could dismiss, so while I waited I took stock of bottom surroundings in some detail, the long range of coal and soot covered pipes in different sizes, that went on to disappear into the distance, large eight inch columns that fed the compressed air into the workings to feed the numerous machinery that depended on air, haulages and pumps and so forth, there were four inch water pipes and pipes of various sizes that ran the length of all the galleries and the labyrinth of tunnels that were supplying the needs of whoever needed them, I noticed the different kinds of cables and the colours that denoted their usefulness, such as black and red intertwined together denoting their use for signalling purposes, like phones and bells, blue for heavy electric purposes, such as electrical haulages, orange and yellow for very heavy machinery, they were all covered in their uniform of dust and soot to which even the strong current of air coming down the shaft could not dislodge, the body of soot that had developed over many years and the same coating of dust prevented the light bulbs from lighting their feeble rays through the mesh that covered them to give the illumination that was needed for pit bottom workmen. I had very often wondered how men could adapt to such a depressing scene that I was witnessing after God's given day light you could barely see the diamond patterned rail tracks that shone faintly under the wheels of the clusters of drams, that crowded pit bottom, what a depressing sight men have to put up with, in the performance of their job. But it was the most important part in the output of coal being kept flowing from the faces, the drams to be placed and directed to their respective tracks t be entered on to the cage, to be lifted quickly to the surface, to be returned to continue their journey in the never ending stream of journeys of drams destined for the loading boys to keep the continuous flow from the conveyor belts. My thoughts as little as they were, were interrupted by the deputy the one I had taken a liking to, and he was in conversation with another deputy who I heard saying "I don't see anything wrong with the kid, he is only like the rest of the kids we have here, and I know bloody well he tries to pull his weight" and I was thinking to myself whatever I do wont please everyone, but the other deputy was saying "orders are orders and that's it I'm afraid" and the voice of my own deputy saying "he will be better off out of the bloody dust anyway and he will live that much longer", so I assumed that my employment at the coal face had come to an end and I would not be continuing my job as a potential collier, not at least at this stage of life anyway, but placed in a new job, what job was I yet to find out in due course, but I would miss some of my friends and characters that I had come to like and understand and I needed all the encouragement I could get, but there you are, you have got to move on I suppose and in my mind I roamed over some of the sayings some of my friends used to say, a chap by the name of Billy Banana, why they called him that I will never know, but that's the name he was known as but I remember saying of the hard times his family had during the depression, he said his mother used to send his brother into next door neighbours and asked them could they borrow the meat from their Sunday roaster to make some gravy and he said my mother used to send my brother to our local shop with a toilet roll and say to the shop keeper, "our mam sent the toilet roll back and could she have five woodbines instead, because our visitor's haven't come", or when they used to go to Barry Island on the miners trips my father used to shout at my brother and say "come over here from there, or we will be taking you home from there lost again". It was witty jokes like that he used to come out with, that you will never forget, because they stuck in your mind, but you will always find someone like that with a sense of humour in that dust and gloom of the pits.

But back to reality again "come on son, follow me, I have to start you on a new job today as you might have guessed, but at least the air will be different than at the face" but like I have said, I have left my friends with a little regret to start my new job, god knows what that was going to be, but wondered what my new venture was going to be like, so we threaded our way through the busy activities of pit bottom which was now in full swing and proceeded in a different direction than we normally take and started to walk up a drift (incline) into unfamiliar smells, that I had not been accustomed to, for instance, the smell of decayed and rotten wood mixed with dried horse droppings was certainly a change to the smell of sweat from your body all day long, but later on I knew the reason for these changes of smells, I was travelling a return airway road where the main air was forced around these tunnels and returned to the surface via the return route, where the foul air was drawn up by a neighbouring up-cast shaft, hence the change of air content, but the air I was entering at that moment was not in any way disagreeable, in fact it was a pleasant change, far from the dust in confined lowness of a coal face and at least you could have a freedom of moving about, and not cramped up in one particular place for seven and half hours, so in one way that horrible puffler had released me from that, but not from thinking about my health but the few extra yards of coal he may get from my absence from the face. Anyway we continued our journey through different ventilation doors that controlled the airflow from the inner workings and came into another passage way that showed some kinds of activity, by the fresh horse droppings which meant that the road was used quite recently by the little pit ponies, delivering the materials to be needed for perhaps repairers of keeping these roads in a state of repair, or perhaps materials that would be used for the coal faces, such as the face supports. The ponies were an alternative to haulage work done by haulage engines that were not accessible by normal transport means, owing to twists and turns in some airway roads, but not all airway roads where supplies of materials were transported. All of a sudden a shout from the deputy "get in the manhole" (a refuge hole placed at intervals in the sides of the roadway to enable travellers to get into cover while the ponies hauled their laden drams or trolleys of whatever that was needed at the time). So I was bundled into one of these little places of safety by the deputy who also joined me and told me to cover my lamp as not to blind the pony as it continued on its way past us. I could feel the vibration under my feet as the pony approached with its burden it came at a fast pace and I marvelled at the way it kept its footings in-between the sleepers of the rail track and the darkness of the tunnel in front of him, given only glimpses of the track ahead by the swinging of the haulers lamp that straddled in the wake of the fast travelling pony, by clinging to the ponies contrivance that linked the ponies harness to the hitching plate of the dram or trolley that the pony was hauling, I could hear now the striking of the ponies shod hooves against some steel sleepers, then they were almost upon us, I could hear the heavy panting of the pony and they were past us in a cloud of disturbed dust created by the churning of its hooves. Looking over the deputy's shoulder I could see the whites of the little pony's eye's as they went by, the hauler had his feet positioned one on the hitching plate of the dram and the other on the curved contrivance that was attached to the harness, the contrivance were called shaft and gun, the shaft encircled the pony's body and the gun was locked into the shaft by a pin, and the other end was attached to the hitching plate of the dram, as they went by I could see the hauler hanging on precariously by his one hand holding on the dram and the other clinging onto the pony's harness, and I dreaded what would happen should the pony fall or he misplace his footing, and then they were gone in a flash. We waited for the dust to settle before we ventured on our way and I thought to myself, the poor little bugger and I thought I was hard done by, compared to him, but that's the way of mining I suppose and I have a lot to learn yet in the years to come should I be still employed at the colliery. We went on our way to a destination unknown to me at the present but I would fine out I suppose in due course, we had travelled a considerable distance I thought by this time and in the distance I thought I saw a light, but it was a repairer doing some alterations to a ventilating door. When we reached the door and it was a large door of about eight feet by eight feet approximately with a square aperture at the centre to regulate the flow of the return air velocity and we had to travel through that door to another passage way, which led to a main transport road, again, I was told to enter a manhole, we waited about five minutes when out of the tunnel ahead of us rushed a full journey of drams from a distant district, from about a quarter of a mile away. It passed us like a streak of white blotches that were on the sides of the rusted drams denoting to what conveyor they had come from, because each district had their own markings. I watched over his shoulder again the shining steel rope that followed in its wake, it was hissing along the ground and sometimes you could see flashes of sparks as the rope made contact with the ground, the hissing went on for a period and finally stopped but the deputy said not to move, even after the rope had stopped and I could see the reason for it after a few minutes. The haulage rope had been transferred to a waiting set of empty drams to be taken back into the workings to be filled again, it was a never ending routine in the circuit of drams in its continuation. I was glad that I was being escorted under the watchful eyes of the deputy and I wondered what I would have done in a moment of panic, when that haulage rope kept whipping from side to side in the narrow confines of sometime twisting rail tracks, but we never moved until the tope had finally stopped and the journey had disappeared from view into the darkness ahead. I followed behind him then I suddenly realised I was walking on familiar ground, it was part of the way I had travelled on my first shift and I could see now why the deputy had made that detour of the airways, not to be in contact with these transport dangers we would have encountered if we had travelled the main transport roads, but we never reached anywhere near the coal face, we had reached our destination well short of that and was where my new job was to begin I assumed. It was at one of the haulage houses that I used to pass on my way out, but I never had taken any notice in the scramble to get to pit bottom and home. I had only taken fleeting glances as I passed but most of the interior of the engine house were covered by brattice cloth, a sort of sacking soaked by tar or something similar, because it had the smell of tar or creosote and I found out the reason soon enough. The reason was it was to stop the cold air of ventilation from entering the engine house but not enough to stop some to circulate but it certainly whistled around it, but when a man has to operate in such conditions for the duration of his shift without getting any exercise it needed that extra protection from that rushing of cold air that passed that engine and you would try and make engine house as windproof as possible and if I was to be given such a job I would also be taking these precautions. The deputy pulled aside the sheeted covering and squeezed inside and beckoned for me to follow in which I needed no encouragement I can tell you and on entering I was confronted by an array of levers and oil covered machinery, levers for brakes, levers for control, clutch blocks to engage the working of separate drums that contained the coiling of haulage ropes, that stretched out in the distance in front, you had the control lever to control the amount of compressed air that you needed to work and move the separate drums for winding to whatever what was on the end of the haulage rope and I thought what have I come into now. It was responsibility now, not just banging away at a block of coal but the responsibility of men that worked at the other end of that rope, it was time for brains not brawn that I was to enter now and had to conquer because I didn't want to make a hash of my second job and I had to prove that I was capable of doing something right for once, and I intended to prove to myself that I was not a complete failure. I was interrupted by the voice of the deputy who was saying tome "well son, this is your new job and I hope everything will turn out all right for you, so now look after yourself and listen to your new friend who will put you on the right road", so off he went on his round of inspection to which he was a little behind after escorting me through the airways where he could have gone on this usual round, my new friend spoke to me then and said "what do I call you kid?" I said Johnnie because I hadn't acquired the nickname of "Chopper" at that time "well" he said, "you won't be doing a lot today Johnnie, so sit down by there on that plank of wood and have a whiff, it won't be long before they will be going for grub anyway". It was strange to be sat here at a haulage engine to which I now knew had existed because I didn't have a lot of time when working on the face, but it was a different kind of environment now, and was free of the coal dust I was used to, the smell of the oil and grease was not disagreeable and the company that I was keeping was a little different from the crows that used to be around me and at least I could stand up and move about. I didn't go outside the engine house, not because of the cold air rushing by, but there was nowhere to go anyway, and it was better to stay in the confines of the engine house in case there was a danger of a runaway dram that has happened in some cases, so I did not see any point in wandering around and it was better to take stock of the different levers and listen to the assortment of signals that was being sent, by the riders of the journey that required the movements of the journey. It was an assortment of signals that required the use of either drum of the engine, there was a sort of signalling that instead of a bell but a pair of ring plates that they normally use to tie the two halves of ring arches together, these were tied together with a bit of wire and they were constantly banging together when the time came for the movement of the journey we were operating and sometimes they were not heard at all by the clatter of an empty journey of drams passing the engine to change the full drams that we had hauled to a point in front of the engine. The empty journey would land on another set of rail track what they called a double parting,

Blue Scars

the rope then transferred to full journey that we had hauled up earlier, the full ones would then be taken away to wait for our next haul of full drams to the same point as before, and we take away the empties that they had just landed. This procedure was repeated daily and that was the work I was destined to do, and I was my own master and I was only governed by the clanging of those ring plates by the side of me, but I had to learn the control yet of the levers but that would come and I felt pretty confident so that's all that matters, but it was more stressful than shovelling coal all day because you had men's lives at risk at the end of that rope, so I studied the use of these levers and listened intently to the instructions of the driver in an effort to come up to his qualifications and once I had attained that, I didn't care a bugger for anybody then, and that is what I had set out to do, but you cannot do that in five minutes can you, so it was up to me to listen and learn and should an error be made then I would be in a position to defend myself, hopefully this would not occur, a mistake could cost a man his life, but there again no one is perfect but in this job you had to be more positive in your approach and that was what I was going to find out as the days passed by finding other potential aspects of driving, by the alignment of your ropes that lead by the revolving of your drums, which would cause the climbing of rope coil against the cheek of the drum resulting in the slipping of the coils into a mess of tangled rope, causing loops and ending in what they call a locked coil and could come in contact with parts, protruding parts of the machinery and including the operator of the engine, so in my own interest I should take notice of such incidents not occurring in any training period, but they say practise makes perfect as long as you don't get over confident but in pit life the absence of daylight and depending on your lamp these irregularities sometimes occur. Anyway in between the blasting of compressed air from the cylinders you have to find time to observe the movement of the signalling plates, when the engine is in motion, bloody hell I thought how am I going to manage all this, but my new friend said it would come natural to me after a few weeks depending on the knowledge I had acquired in the weeks that he stayed there. Take notice and don't crowd me and hinder him when I am rushing from one brake to the other, so I found a good vantage point where I could observe without getting in his way, but it was nerve racking to me as a young boy to learn so quickly and to show you were capable of the job and I intended not to fail because of my failure at the coal face, so after the deputy had gone and we were on our own he said I am glad of a bit of company to break the monotony of his shift, so my first job was to oil the moving parts of the engine before he commenced his start for the day. So one day when the twenty minutes grub came around he said to me "come on Johnnie, I will show the roads to you so you will know how to control the journey with the undulation in the roadway so while we were doing that we visited the loading point where the coal conveyor was delivering the coal into the drams, our journey was under the loading belt in readiness to when we were needed to haul the next dram under it, one of the men shouted at me jokingly "cause us any trouble you little bugger, we will put you in these drams and fill you out with the bloody coal, do you hear me?" I took the hint as a sort of warning as we started our way back out to the engine but my butty said "don't take any heed of him he was only pulling your leg". Anyway one day when the coal was slack in coming from the coal face my butty said to me "next time they give the signal, have a go". I did but I had forgotten my rope that was holding my journey together, they called that the tail rope, the main rope was the one which hauled the journey outwards and I hadn't kept my foot on that tail rope resulting in the journey jumping forward. Bloody hell I thought, now I am for it, I could hear the men shouting in the distance and remember what the chap had said and I could see the lights bobbing about. One of the men came up to the engine, my butty said it was his fault but the men knew different because of the routine they had been used to, so the man was not nasty in anyway and was quite tidy in talking to me, "try to remember next time son, I know this is new to you but it is a good job that there isn't a lot of coal coming at this present time or there would be a mess of spillage that would need to be cleaned up before they could start again, so be careful next time. I know we all have got to start sometime I suppose". I said I was sorry and all he said was "don't worry kid, you have murdered any bugger but be a bit more careful". I thanked my butty for trying to take the blame and I carried on after I had made that mishap and settled down to the job in hand and my first hurdle was to bring a full journey of coal to its resting place which was more or less in front of the engine and rested it on the monkey (Jack Ketch) and take the empty journey back to the loading point without any more trouble at all, thank god. I carried on and my confidence returned and was less stressful so by the time my butty had to move from there to a larger engine outside of us I was more or less a competent driver and during grub times I used to stroll down to the loading head and have a chat to the fillers, two of them were getting on in age a bit but like everything that happens to a new start they wanted to know who I was and where I came from, who my father was and wanted to know the ins and outs of everything, that is what it's like down a small pit which was more or less a family pit where everyone knew each other. They were joined by other men who worked in the proximity of one and other by men that had the job of oiling certain machinery which needed constant attention, but you was always put at ease by some of the jocular remarks made in conversations and sometimes a bit of horseplay would come into it, but all in all they were a decent bunch of men to be working with. Some had a habit of fooling about with different objects that lay about the place, like putting some bodies shovel in a dram which eventually got filled out in a dram of coal, but there was one thing that men never played about with and that was mates food and water—that was taboo and was completely out of order to their code and conduct and some would say yarns that were probably lies anyway and some jokes that were not meant for my ears and were oblivious to my presence. I did not know what they were on about anyways but I would not like to repeat, but some were laughable jokes like one old timer said after spitting his bacca out of his mouth "my mother used to send me down the butcher shop in the middle of the week to buy a pig's head to make stew and pigs head brawn but I was to tell the butcher, could you cut it as near the tail as you can" but another oldster by the side of him said "it must have been the same butcher our mam had because I had to say to him, our mam wants a pig's head to make some broth but could you leave the back legs on please". These sayings went on and on during grub time, one or another trying to best the other, but I was loath to get back on the engine to start up again, but they would still carry on I expect after I left but it was a ray of sunshine to find that even in the gloom of the pits there was that sense of humour to be found, and in discovering their ways while working with them in later life (should I continue in my mining career) you don't feel as isolated and you will find the true value of the relationships that develop over the years to come, that creates a bond of friendship and trust that only miners could give you during the close proximity of working together and I think the memories of this will stay in one's mind throughout your life.

So I made my way back to my engine, a bit reluctant after grub time but hoped in my innermost thoughts that there would be a breakdown of some sorts in the machinery to allow one to continue mixing with that group at the loading head so when one of the men said "come on son get off your arse and get cracking and let's get started" I was already on my feet and on the way back, my butty said to me "I have heard those same bloody stories over and over again, but they said it for your benefit to make you feel at home, but woe betide you if you make a mistake I can tell you, they will skin you alive, but I can see that you have the hang of it now Johnnie". I said you can always learnt something butty, but later on the loading stage was moving inwards as the face advanced and that put me at a disadvantage because doing the procedure I had been doing for a few weeks was altered by the fact that the length of the rope had been increased by the extension of the loading stage, hence the chalk marks that I had put on my ropes that gave me the position of the journeys to a fine art were completely altered. I know that rope marks on ropes was against regulations but there was not many engine drivers that did not have the same procedures because it put them in the position of knowing to a point of being accurate to within a foot or two of their mark of which myself had been accustomed to, so on one occasion I was confronted with this problem after a few mishaps which the men at the loading stage made allowance for till I had adapted to my new routine, but I did a few miscalculations such as not allowing for the extra convolutions in the road in advance of the loading stage, but I managed to overcome them as time passed, after a few little confrontations with the men at the loading stage but I had not put any lives at risk, and put it down as an occupational hazard, that could occur to any driver and I had to put up with a few expletives of language that I never heard of in my life, but I knew what the message meant to me, but anyway things got back to normal as the time went by so it was back to the usual routine and that episode passed and work resumed as normal and I dreaded the next advance of advancement of the coaling stage, but I could not afford to keep on making errors but my driving skill improved and I overcame the biggest part of my problems and dealt with accordingly to every bodies satisfaction thank god.

I knew the biggest part of the men that worked in and around that part of the district and they were hardened miners from my point of view, but people have different appetites for work and different levels of stamina and I was of small stature tried to cope with the latter, while my appetite for work was in its infancy but nevertheless a challenge to what I was most suited for and the uncertainty of what transpired out of my apprenticeship and what selection of jobs that was required of me, but I must have weathered the storm because I was there for forty two years to which I have already stated, which shows that I was capable of both appetite and stamina but to identify what really attracted me to mining was the close relationships I had with fellow workers who with their support and encouragement perhaps I would never have survived, but I never regretted a moment, but I surprised myself by making a career out of it, but on realising the part I had taken, during that time, I wish I was back there, but owing to age and the demise of the mines know it was impossible, all I can do now is reminisce over memories but in time even those may fade away, this is why I have decided to write my biography of my life in an attempt to make a social document that will I hope stand the test of time, this may seem boring to some reading it but it is a testimony to a forgotten era for generations to come and compare the uncertainty to what the future may lead to.

I have moved now to a larger haulage engine quite a distance from the loading bay when another boy my age apparently had taken over and I wished him the best of luck and given him a few tips of the running of the place. I can imagine the amount of frustrations he was likely to get that I had when I started so I knew what he was going through but I myself had to get used to my new job but having had experience of that type of work gave me the advantage of coping with a larger haulage engine. I used to go in a little earlier in the beginning to get the feel of the new road of which they were much longer and catered for a larger length of journeys where as the smaller engine coped with the amount of drams they were subjected to, almost half the length of what I was to haul, but I did have the advantage of a signal bell instead of a bunch of ring plates thank god, but the road that I had to cope with was the gradients and the undulation that I had to master, that is why I went in earlier to size it all up and to feel the shift of weight as the journey entered these undulations in the particular areas, and to obviate the difficulties that may occur, namely derailment of drams, but in that event, the rider of the journey and I had a code of signals that would apply to what movement was expected of me, such as the signal of three to haul, the last note of three was at a longer interval to which I interpreted as ease my journey forward very slowly, the offending dram or drams, it was not a problem for me and it was a workable system to me and journey rider. I worked with a different set of men now, but with different ideas, some with wisdom and some with live today and bugger tomorrow but all dedicated to the job at hand and as usual made my life tolerable. Some acquired different nicknames denoting their jobs, and some with their mannerisms, such as Dai Baca because he was always cadging bacca as soon as the cage landed at the bottom of the pit and saying anybody got any bacca to spare because it will only go dry I the baths, or he would say bloody hell I have left my bacca in my clean clothes, but the excuses would not wash with all of the men, who knew it was old Dai's habit, but still insisted on his usual jaunt into his work but it would be a poor atmosphere at work if you didn't have blokes like Dai to give their contribution in making a pit more tolerable, but whatever industry men or women worked in there was always one or two characters to be found, but more so down where I worked thank god for those characters, we had another character to add to our selection, a man called Billy Dusty who was employed as cleaning the rail track of debris that cluttered the track of spillage caused by ill fitting doors of some drams, he was nights regular because he said he could not work days or afternoons because he had pigeon lofts to look after, so he was allowed to work nights for that reason, or perhaps he did not have enough of vim in him to do anything but that job, I don't know, or was it craftiness, we will never know, anyway that was his night job, well one night he must have been buggered up from staying up all day to see to his pigeons, that one night he must have fallen asleep while having his grub in one of the manholes by the side of the roadways,

Blue Scars

which was a serious offence to fall asleep underground, well on a few occasions men were deployed in spraying the areas of roadway with white lime to combat the dust collecting on the sides of the roadway, and that would also add a bit to the illumination of these roads, plus these men had to scatter stone dust about the place, like I said to combat the dust that was everywhere, and was a deterrent to avoid the risk of an explosion caused by the dust that may have an explosive content and stop the explosion from spreading, because it was imperative that this sort of work was carried out in the interest of safety but Billie's job was to clean the rail track and empty drams were placed at various points for Billie to do his job and like I was saying he must have fallen asleep. The lime sprayers were getting closer to Billie's manhole and as they approached they saw that he was still asleep, they did not wake him up but sprayed his manhole with Billie still inside, but he still did not move, so they gave him a second coating which was drying rapidly, they sprayed the shovel he was holding between his legs and gave him the final touch and then went on their way, down the roadway, but if there was an official in the vicinity they would undoubtedly have awakened him I suppose, not to get him the sack, but eventually they had to leave to continue their work in the next roadways, the story goes that when Billie woke up the lime had set like cement and if he had walked that roadway and met another workman it would have scared the living daylights out of him, thinking it was a ghost, but they say he had a job to free himself from his shovel and was walking like a stiff legged zombie, but Billie never slept in that manhole after that or any other place and he could not complain to anyone about the incident that was why he was called Billy Dusty. It was a few pranks if you can call them pranks as the case of an electric haulage driver, they say idle hands can make mischief well they have different levels of coal seams called horizons, but on one level called the black vein an electrical haulage was situated in an airway, behind the haulage was a breeze brick wall and by the side of the wall was a ventilating door, which held back the air but was allowing enough through a slot in the door. But to ventilate the haulage sufficiently a six inch pipe was installed in the wall behind him in which the air blew quite strongly, the pranksters were aware of this and while waiting for a supply to pass through the door, well in most every pit the pony stables was always occupied by the stable cat who had a good living off the mice that scuttled after the feed of oats or any other form of feed that the ponies required but one of the men entered the stable and caught the cat and carried him to the pipe in the wall while his butty covered the hole, after the cat was placed inside and waited for the pressure to build up against the board he had placed against the pipe, so when the board had enough pressure behind it they whipped the board away resulting in the cat being blown through the hole and into the engine house and flew past the unsuspecting driver, who on seeing this apparition flying past him was petrified at the sight of a flying cat and his shout you could hear from the other side of the wall, but that prank was never repeated thank god and was never reported, the only comment the driver made when the subject arose was what if the engine was in operation at the time he would have lost control, all he said if I had caught the buggers at that moment they would have needed hospital treatment, that's the way I felt at the time, it frightened me to bloody death.

Back to my early days at Beynons, where one incident will stay with me the rest of my life, it still is, and I can remember it to the present day, I was in my early teens, eighteen or nineteen at the most and I was waiting for the men to pass my engine to go onto their work, before I could start operating the engine, the empty journey was standing at the side of the engine at the time and my butty the journey rider had gone on ahead to clear the track should there be any obstruction on the track that would derail us when the journey was started on its way. It was the usual procedure he did everyday and when the men had disappeared into their workplace, he would then signal me to proceed with my journey. So I sat down and waited for that signal when an elderly gentleman came into the engine house and said he had placed his tools in the last empty dram of the journey, I said "right o" and waited for my signal to come, it came and I started to move my journey inward when the last dram appeared by my side. I stopped and said to the chap unload your tools and I will drop the journey back and you could place your tools in the front dram, because it would save him carrying his tools past all the empties in front of it because it is narrow in parts of the roadway. I thought like any sensible person he would have thanked me for helping him out but on the contrary it was the opposite of thanks, he replied in an un-friendly way of talking to me "to mind my own bloody business, I will manage without you telling me what to do". I just looked at him, I couldn't believe my ears at what he had said, so I said "alright butty I'll do what you say then" and as I started to move the journey he said "hold it a bloody minute, I am having a ride in with my tools". I said it was dangerous to ride in the journey because when I reach the curve in the road the last dram would tilt owing to the braking around the curve because of the weight of the drams in front and I was only trying to help you Sir, but it was the way he spoke to me in a sort of bossy way and

I was to listen to what he was going to tell me. I looked over and I thought to myself he was not going to do a lot of work attired in those clothes he wore but said only to him that it was dangerous and against the rules and I was the one responsible if anything happens. So I bit my tongue but I was seething inside when he replied "mind your own business you little b—d and get on with it", so I did and by the time the journey was half way to its destination I was boiling over for what he had called me and I increased the speed of the journey and didn't care what happened to the last dram and by that time I was doing a decent lick but it was too late by the time I had reached the curve and had to brake heavy, resulting in what I knew what was going to happen, but my butty did not know about this because he was in front and stood at our destination point so he could not see what happened beyond the curve so he I stopped at his signal and pegged down my brake and waited but in my own mind that those last two drams would be closer to tipping if they had not had the bulk of the brickwork to stop it, it would otherwise have turned over but I didn't seemed to have cared after he had provoked such anger, but it was out of my system now, it was too late, so that was that. I would certainly get the sack by not controlling my temper, when a light appeared in the distance, it was my butty, he said "what in hell happened, you wouldn't stop when I signalled and I thought the signal bell was not working, and you were coming in at a speed not normal to the way you drive and I thought something had gone wrong", so I told him the truth about losing my temper when he called me a b—d for trying to help him. "I can see that now but you should have seen those stream of sparks coming off that brickwork", I said "how could you see it you was supposed to be inside", he said, "I didn't have much bloody time, all I could see was a face in the last dram so I went and had a look who it was, the man was mixed up with tools everywhere I know it's not a good time to laugh Johnnie, but it was the way the bloke looked at me, when he tried to get out from that dram, that is what was most amusing, he didn't know where the hell he was, all he was saying to himself and cursing you, and I thought to myself God help you when he catches hold of you" I said I didn't worry and didn't care what happened to him after calling me that name, but after I had cooled down, I realised what I could have caused but it was too late now anyway to relent, but what was going through my mind was the thought of the points of the picks and that sharp edge of the hatchet, I could imagine how terrified that man was at that moment, when the dram tipped on its side and I would not mind betting that he would not be riding in any more journeys after that episode, but I consoled myself that he was the one that drove me to it, by all accounts my butty had asked him was he alright but the man just stared at him and wandered around dazed. My butty said I couldn't get inside fast enough to tell the men what has transpired but when he came out with the next journey the man said that after he came out after his shift he would go straight to the manager to report that I had tried to kill him, to which he wasn't far from the truth, I suppose but the thought of really doing that was not far from my mind, but what I mostly wanted to do was to frighten him out of his skin, but like I said it was too late now because the deed was done and all my butty said was "you had better watch your step John when he comes out". I cannot avoid him very well because he has to pass me to go out I said, but anyway by the look of his fine clothes he was wearing he had no intention of doing any bloody work, the tools he was using had no sign of wear, so he was only there to collect a day's pay, but I don't think there was any work in him, nor had the intention of grafting. You can tell some men by their attitude and mouth of which he had plenty of, but eventually the shift came to an end and the men filed past me on their way out and one of them shouted to me (laughingly) "what have you done again you little bugger, you will be for it when gentleman Jim comes out and he aint for being kind, so watch it kid", but as an afterthought he said "serve him bloody right for calling you that Jonnie and bugger him" and off he went, but it was alright for him to say that because I was going to be the recipient of that guys anger and face the music, and I was a bit apprehensive, not of the man, but what he was going to say in the office and he had all that day to build his case whereas I had no defence anyway. But it wasn't any good worrying about it, what's done was done and that's the end of it, that's the worst of being a bit headstrong at times and there was no lack of it by myself although I cannot remember when I lost my temper last, because I never had cause to because there were so many good friends around me, but this chap sure hit a sore spot in my armoury. So by now most of the colliers had passed me and I spotted him in the distance followed by my butty, he was getting nearer and nearer to the engine house until finally he reached me and had to pass close by owing to the journey by the side of me, naturally I took stock of him because he was as clean as when he had entered the district so I had sized him up for what he was, he hadn't come to work by the look of him but to talk his way out of it, he had the air about himself of being upper class and the little people had to bow down to him, so he must have had a bit of a shock by being in that last dram by showing he was just as vulnerable as the men he worked with, anyway he was passing by me but I had a shock myself because he looked

Blue Scars

straight ahead and never opened his mouth and never gave me a glance. I heard the men behind him sniggering by saying to him "why don't you talk to your friend Jonnie", but he never said a dickey bird, but I thought no he was saving all that up for the manager when he went up the pit, and I knew my butty could not defend me, because he wasn't there, so it was one to one, his word against mine, and I can tell you, my word was pretty weak, all I knew I had a good friend in my butty who was more like a brother than a workmate. His name was Cyril Randle who lived a few houses above the Queens public house in Abertillery road, Blaina, he was a good man, he had a nickname of Duck Randle a west side footballer of Blaina, I will always remember him for his good nature and likeable ways, I cannot say enough about him, but sadly after I left Beynons Colliery he was killed by the very drams that he was so used to handling, but that was after I left Beynon.

But back to the present and my predicament I had put myself into, so when the men had all gone by and out of the way I pegged down my brakes, shut off all the blast (compressed air) collected my blocks of wood I had cut off to take home to cut up for the fire in the morning, because men were allowed the block of wood by management which was one of the perks used in the pits, providing they were eight inches long and were cut off used timbers. So I followed the men to pit bottom to wait in the usual queue and waited to ascend the shaft, but by that time my little episode had preceded me and a few glances were towards the toff, there were sneers thrown at him but he never said a word, and the feeling of some of the chaps that he was not liked at all, so they must have come in contact with him beforehand apparently but it did not alter the fact that I had done wrong and had to pay the penalty for it. There was no doubt about that so I accepted it, my father knew nothing about this because he was already up the pit bathed and on his way home by now because haulers were down a half an hour before us to attend their ponies, and prepare them for their daily toil, but I expected he would know in due course. So my turn came to ascend the shaft and as usual handed my lamp into the lamp room and was told by the lamp man that I could not raise my lamp the next day until I had seen the manager to which the news was expected and did not come as a surprise to me anyway, but I thought that was bloody quick by any standards, the bloke did not waste any time by the look of things, reporting me so I took my time, went in and had a bath and went on home with a few of the lads who were laughing at what I had done, but I told them it was a laughing matter but the answers that they came back with, was exactly what I had done to the bigotry bugger, had the occasions happened to them, so when I arrived home I did not mention the incident to my father but would know soon enough I expect. All I know is I didn't sleep a lot that night I can tell you because I was worried sick by now and now the reaction was showing itself, but I did manage to grab a few hours sleep during the night, and was buggered up by the time the men called on their way to work, so off I went to work and arrived at the pit and was greeted by some of the colliers having a fag before they went under "best of luck Johnnie" they said but that did little to comfort me, because I had the same feeling in my stomach to when I met the manager at first starting at the colliery and as I waited a few deputies (officials) were emerging from the manager's office after receiving their orders for the day, and I could see the under manager by the grimed old windows, so after the last of the deputies had left the office, I was left to stew for a bit in anticipation of my interview, but finally it came, I thought bloody hell now I'm for it, the manager who was at the window at the time beckoned for me to enter which I did reluctantly, not knowing what I was about to face by the under manager and the manager. The manager's name was Mr. Tom Bowen and the under manager's name was Mr. Powell, a stern looking man and when he finally spoke it was very quietly spoken and after standing by the window for a few minutes he turned around and said to me "we have a very serious complaint about you Johnnie, which does not look too good at your place of work, and you answers will determine your future at the colliery Johnny do you understand?" I replied meekly "yes sir". "Well son we will proceed with your explanation of what happened, your father wanted to be present but we said we prefer you to be on your own and that you are quite capable at your age to be judged on your own merits, do you understand". I replied "yes sir" again. "We have heard the complainant's side of the story and now we would like to hear yours to be fair to both sides". I said I understood, "well let's hear your version son, I don't want lies, but tell me in your own way and don't leave anything out, so proceed son and this may depend on your future with us". So off I spoke in a more firm voice that I expected of myself. "Well everything was working as normal like it is every day sir, till that man came into the engine house and told me that he had put his tools in the last dram of my journey, and all I said to him was in trying to help him was by telling him it would be better for him if he placed his tools in the front dram, thus he would not have to carry them past all the empties as it was narrow in some parts of the road, you know where I mean sir?", "yes I understand what you mean son, go on", well I continued with my version of accounts that happened "well sir the man told me to mind my own business in no uncertain way either", the manager said "did he at anytime swear at you", I said "no sir, not till later till I had pulled my journey up till the last dram was by my side of the engine", "what happened then son?", well I said "he shouted at me to stop which I did then he said he was getting into the dram with his tools and have a ride in", all I said to him was that he was not supposed to ride in the drams, that was the rules, and not only that I said it was dangerous to ride in the last dram anyway because the weight of the drams in front of the journey would cause the last dram to tilt by me braking, the man said to me "do as you are bloody told you little b—d". I told the manager full mouth the word he used to me, "I know I should not have lost my temper sir, but it happened and that's all I can tell you but I can imagine the dram tilting on the curve but that was expected sir". I didn't tell the manager what really I was thinking at the time so I concluded by saying I have told you the truth and that's all I can say sir. The manager spoke then "do you know that this man wants you charged with attempted murder and that's very serious son and that man is a well respected councillor at Brynmawr. I can name the man but it would not be in the best interest to name him so I will not do so, I am only interested in what happened, so what shall we do son?" I said I didn't know but all I knew is he provoked me by calling me that name and he had no business to ride in the journey that's all I know sir. "alright son" the manager said, "go outside till we are ready to call you in to make our decision", so I waited for a decent time, I thought, but the moment came for me to enter the office, they looked at me and then at each other and said with I thought a hint of a smile, not very much mind you, but it was there, I knew it was not a stern view as when I entered the office for the interview but enough to know that I may

Blue Scars

be let off the hook. He said you both broke the law, you yourself by putting life at risk by losing your temper and the other man by breaking the law by travelling in the way he did after being warned of the consequences of doing so but remember that temper is not needed at a job that could result in serious accidents do you understand son. I replied meekly "yes sir", "so off you go and remember what has been said, go and change and get your lamp and let this be a lesson to you my son", so I was back in service so to speak, I went and changed into my working clothes, then collected my lamps but it was too late now to go down the pit because coaling had started winding so I went to the canteen and had a cup of tea with some of the men off the night shift, who was waiting for their buses. I didn't have any money so I had to stop it, but after hearing of my ordeal at the manager's hands I didn't have to pay for my tea, not only that, I had a nice piece of cake to go with it, so after I had eaten I made my way to the pit top and waited for a break in the coaling system and when that time came I descended the shaft quite alone in the cage getting my thoughts together, still remembering my ordeal in the office, but that's all over now thank god, but it taught me to watch my step in the future, when given a second chance, so after landing at the bottom I did not hurry because half the morning had gone by so I wasn't worried I still get paid anyway. I made my way to my destination the engine house to be greeted by the chap that was filling in my job while I was away and did not know if I would be returning but he was later informed by the deputy that I was on my way, and when I arrived my substitute informed me that it was too bloody late now for him to go back into the face so he said to me go and cut a few blocks for the boys and have a whiff. It was a novelty to me to see someone doing my job but I didn't mind I was getting paid so why worry about it. I asked him had he seen my friend, or shall I say foe, he replied I dont think he is in the district because his tools was on one of the journey of coal I pulled out this morning, so he won't be back I shouldn't think. A bloody good riddance I say, so grub time came and my butty came out with the journey and I gave him word for word what went on in the office and my reprimand that I faced, and while we were having grub the deputy passed by on his round of communication and had word of my reinstatement and also heard that he was to apologise to me, but that never came about, because the man never came into our district again, at least not that I knew of anyway and that was made more conspicuous by his absence, and never had any more contact with the thing, other than a part of his tools came out on the journey, but I was put on a sort of probationary period by management and that soon faded away after a few days but I knew that the men in black gave the toff short shrift in the short period that he was in the district, not because of his councillor status because there was not a mark on his face or likely to be, in which he was not, they judged a man that was not afraid of work because mining men hated scroungers of that sort, but he was gone now, and that episode is finished thank god, except for a few comments from the deputy, who in conversation said that I had not pulled the wool over the manager's eyes.

I stayed at Beynons Colliery for about ten years and decided to try outside work to try and broaden my outlook on life of everyday, but found that mining was still in my blood by now and not only that, I found out that outside work as little as I saw of it, did not come up to my expectations because to me I had entered a world of what I saw of it was a proper rat race and quite a lot of back stabbing going on unlike the world of a miner I had just left, but while I was out of the pits for a couple of months I fell into the trap of being called up for national service to which I did not foresee but was discharged after eight months for not fulfilling army physical requirements namely eyesight, of all things, but I was very relieved at being discharged, because by that time I was twenty two years of age and was married with a baby son, so with that responsibility I was desperate to find a job to support my family if I wanted to give them the same life as everyone else. That was the year of 1945 and I was discharged from the army in 1946 so one day after a conversation with my brother-in-law decided to try and find employment at another colliery, which by chance happened to be at Six Bells colliery Abertillery and luckily we both had an interview because there were a few vacancies there, but on arriving found out that there was only a vacancy for one of us, but I could not very well accept it, because the two of us had tramped miles looking for jobs and I could not accept on principle but after explaining to management that we were a team (to which we were not) that knowledge of pit work, and that we understood from sources that the vacancies were for conveyor fitters, and after a few white lies said we were conversant with that type of work (again to which we were not), was finally accepted to be part of a team, but I refrained from telling them that I had failed an army eye test, that's how desperate the situation was at the time, and after a bit of questioning of some aspects of mining, to which I did have ten years but only as a face worker and a haulage driver, but it apparently satisfied the official that the interview was with, was accepted and we were signed on as conveyor shifters, to which we had no knowledge of but it was settled, and that was the start of my thirty odd years of the pits, to which I never regretted and gave that incentive to build a home for my family with the help of a good mother and father-in-law, and the best wife in the world, whose name was Tilly Phillips who lived in the same street in Vincent Avenue Nantyglo and we had a lot in common, and knew of hard times, but I had better get back to mining now or I will get carried away again by reminiscing to which I could relation in another hundred or more pages.

So my brother-in-law and I completed our interview and was told to report to the lamp room to put us on the register to claim our lamps when the time arrived and that was for the following week of the afternoon shift and this was my second attempt of being a miner and far from being a haulage driver, to which was to find out, but not deter me, from getting back to the job I loved doing and that was the main thing but my first responsibility was to my family, it was my first consideration also to the manager that gave me the chance to start again which lasted over thirty odd years in the same colliery and the starting of a story in those claustrophobic surroundings and I have poignant memories of some stomach churning events that I encountered during that time during a lifetime in the pits, not always occurring at the coal face, but in various tasks men have undertaken, in the pursuance of those black diamonds and to keep the coal flowing for the employment of its work force. I am not trying to paint a picture of doom and gloom but illustrate the dangers of extracting that mineral and the wisdom and characters of some, that defies those dangers including those acts of heroism that go unseen in those dark corridors, but we must not speak harshly of the pioneers of the past and the communities that were to grow by their efforts and that is why I pay my tribute to such men, that did not have the opportunity to use the technology that are available today but the slow demise of the pits today have not dampened spirits but made them more determined to fight for its future, some men are born to be leaders where others live for today but in all cases men have to abide by the rules and failure to conform to these results could result in the failure of the system that is there to protect them. Safety is paramount and a motto is take care and beware.

So I began my shift with Dai my brother-in-law, we started on the Monday the following week and were shown to the officials cabin which was situated near the pit bottom which is similar in all pits, but the similarity ended by a larger pit bottom, but the same dome of arching that resembles a huge Nissan hut, the same array of endless pipes and cables that was the make up in all collieries, you still sense the feeling of comradeship that only a mining workforce can bring out, familiar greetings of "come to have a go have you boys?" and the usual answer of "try to anyway butty". It made you feel you were back home, but my main concern or shall I say our concern was how we were going to familiarise with the challenges of the machinery we were about to be introduced to but fortunate for us was the fact we were only needed as replacements, to when one of the team went absent, whether through and accident or had a day off to which we would fill in that post, but if that happened there were always the remaining team to teach you as you went along, if hat did happen, then it would be the deputy's choice of who he picks to fill the job at hand, out of Dai or I, but it was too early to say when that would come about, meanwhile we were given jobs of a sundry nature, a sort of on loan to anyone that needed you and that was mostly what our jobs were in the meantime, but we were not paid on piece or contract work such as face work or conveyor shifting, but paid as an ordinary work man but it was better than dole money and it was a regular job, so we could not complain as to what job they required us to do, it suited me to be back amongst the soot and dust that I was always used to, so every morning or afternoon we were to wait for our instructions to be given us at the officials station and went on separate jobs, I don't know where Dai used to go but on occasions we did meet. My instructions on one day were to help out on a supply journey which was to take supplies of material to various districts and sometimes helped to load timbers and ring arches into various assortment of drams and trolleys that catered for different lengths of timber and ring arches, and some trolleys to accommodate lengths of pipes that were need to extend the often needed pipes column near the road head near the coal face, these jobs were new to me because of the transport system of Beynons, so half the shift was in the loading of miscellaneous mixture of materials to be placed as near to the coal face opposite to get it, but your two butties to work with, one was the engine driver and the other doing the same as myself, but I was on unfamiliar territory to that of Beynons, different districts, different direction of roadway, but most of it was left to the regular driver of the supply journey, while I just followed his instructions of when to do the signalling part, or when he was not in a position to handle an item of ring arch by himself and needed my assistance, well that was the idea of being there on the job in the first place, but it worked out that we both needed one another so when we landed our journey of supplies I would signal the engine driver to come and give a hand if needed and that was teamwork, after replenishing the needs of the road ahead and coal face we used to have a whiff for ten minutes to come to ourselves and the driver returned to his engine after sheaving the rope around the curves in the roadway to help the first dram around the curve, he had the arrangement of touching the rope with a little tug to let us know he was ready to haul the empty supply journey back out to our place of loading, but not before I had jumped in a trolley to have a ride out and not to follow the journey because sometimes the rope steadying the journey behind used to lash from side to side of the narrow parts of the tunnel, but we always kept an eye open on the leading dram in case it got derailed at the curve of the roadway, and we were always in readiness to grasp the signalling wires that lined the sides of the roadway, it was not a problem, but we were always anticipating should anything arise on our way out but we mostly got to breaking the law by riding out in the empty trolleys and I more than anyone else should know what the end result would be by that episode in Beynon years ago and the boot was on the other foot now, but it was the lesser of two evils, it was either that or having an accident by that swishing of that rope behind so I chose the safest way and the supply journey was not a hundred miles an hour run anyway. But the next time we had a journey of supplies was a different kettle of fish it involved the transporting of lengths of pipes that needed constant attention owing to the swinging of the trolley that contained them, the trolley was more or less isolated on its own because of the long chain that coupled the journey together, you could not couple the drams and trolleys together by the usual one shackle (coupling) because the vehicle loaded with pipes needed that extra length provided by the long chains to manoeuvre freely around the curves, even if you had them last of all on the journey, the pull of the rope would act as a pivot so it was in our interest to place the offending trolley of pipes in that position and not too far ahead, it was different to that of timber which had an average length of 8 or 9 feet which would allow shorter lengths of chain couplings but with pipes you had an assortment of lengths varying from 10 or 13 feet in length and woe betide you if by chance that the swaying trolley of pipes was derailed, it was hells work to re-rail it, so really you had to depend on which way the driver got out of bed that morning, I know by experience of having been an engine driver myself but that was mostly on coal transport and not in these little narrow passages that needed a little more experience in negotiating, but every man to his own job I suppose. I used to marvel at the tactics used by my former friends when re-railing the drams by using pieces of short rails, preferring the use of latch rails (points) which had a tapering end and that would allow the drams to enter the grooves of the rail and guide them up onto the rail track kept by placing half bricks or anything that was available underneath the rails to steady the tipping another rail similar, which was placed parallel to each other. It proved effective in many cases but depended on the skill of the engine driver to ease the journey inch by inch up these improvised ramps, but anyway at the present job of supplies many time I used to ride the rope and steady myself by holding onto the dram giving me a better view of the rest of the journey in front of me and thus avoiding some of the cavities between the sleepers, not very ethical I suppose but it was most convenient because those cavities or holes were there for a purpose also, by allowing the little pit ponies more height and frequently travelled the roads in the absence of the journey that were not needed just for a single item, that was needed in the shortest time possible. Back to the journey now, the engine rider had a special code of

Blue Scars

signals when difficulties arose at one time or another and that he had to visualise what the problem was, not seeing his journey a hundred or more yards away, but it came automatic with years of training and the knowledge of the rail track, but in time there may be a more rational approach to the way of delivering supplies and there was room for improvement in the current practice, well to move on after there were adequate stocks of materials and the journey was not needed for a few days. So my next job was in complete contrast from being a supply man, I was given the job of helping some repairers handling ring arches by steadying the one half of ring section while his mate steadied the other half and when the two halves matched together the other of the team bolted the two together with plates, then it began the work of building behind the ring arch, to steady the entire arch but filling and packing with dry walling interlaced with four feet long lengths of timber, the cavity left behind by the walling were filled with debris from the roadway, thus consigning the ring sections from the weight of the side squeeze when the ground settled eventually, after the work was very near completed I caught hold of that hatchet to trim a length of timber but if I did there was a shout "oi, kid, leave that bloody hatchet alone, only I use that and you would miss and hit a lump of stone and take a chunk off the bloody blade, so leave things alone do you hear", so I put the hatchet down and should have known better because I remembered the oldsters from the coal face because they need their hatchet as the main thing in their armoury and it was taboo to use another man's hatchet without their permission, but to get their permission was another thing, I know, because I used to watch them avidly the preparation they put into their hatchet, by some of the oldsters, they put into their tools at any five minutes that were available, they placed it between their knees and with the aid of their sharpening stone had the requirement that they needed in a razor like edge and I have caught the withering look that was given to anyone that asked for the loan, because without it they could not proceed with the work of cutting supports to cover their backs, so I could understand the way the chap shouted at me, so I carried on building the wall to shoulder height and some of the wet timbers were heavy by the height at which they were lifted, but the annoying part for me was the resin that seeped from the timbers which took days to get rid of because the glue like substance was mixed with dust and coal and made patches of black all over your hands and arms, but after days of coming in contact with stone dust occasionally rubbed into the affected area they would eventually peel off, but you could not rub off the cuts and abrasions that your hands had by handling sharp pieces of stone and rock, but at the end of the day I could hardly pick my arms up and with a stiff neck from lifting I was walking like a wooden man, this was the only thing of being a spare hand, there was not a fixed pattern to a week's work because you were here one minute and the next minute you could be half a mile away to another district where you were expected to fill in a vacancy in unfamiliar working practices, but there is no substitute for experiences but sometimes these experiences are a clear form of education but knowledge may be a long way ahead and sometimes ones time is short and far too quick but my experiences from my former pit gave me the advantage to overcome most of my problems but still you can always learn, but there is no end to the pitfalls you may encounter during the period of employment in the mines. So the next day I was placed on supplies but only to sort out the stock pile of sundries that littered the supply heading of the road ahead and putting

Blue Scars

So some days to break the monotony I used to change jobs with the engine driver, he does the rope job and I do the driving, it was terms that were agreeable to both of us, so on that basis we carried on and I was in my oils so to speak. It turned out to be a pleasant arrangement.

I am back at Beynons Colliery where the tricks and pranks started as a young hopefuls entry into the underground city, there were a few wicked characters in my district at that time but not in the vicious sense of the word but bloody aggravating, and tricks that were played on unsuspecting new entrants to the pits, but as a kid I remember some, on one occasion an oldster working at the coal face said to me at grub time "Johnnie when you come to work in the morning do me a favour", he acted his part well, not smiling and neither was there any indication shown on the faces of the chaps that were sat around" "well Johnnie would you call in the stores and ask the store man could you have a tin of stripe paint as we want to pain the supports in the face so that we can see them better, but make sure that you are not late to catch the cage with your butty". I said I would, so I tried to remember in the morning what the errand was, so in the morning I made enough time to call at the stores, I knocked at the door of the store room and walked in the, the store man looked at me and said "have you come for chalk for the drams", "no sir" I said, "the men of our face told me to ask you could they have a tin of stripe paint so they can see the timber at the face better", he said smiling "sorry my son, I haven't any stripe paint spare but if you could carry them out I will paint them for you", so I went down the pit and reported to the man what the store man had said and he said "doesn't matter son, we will try and manage somehow", and he didn't not even smile. I thought to myself how could they carry all that wood up the pit, one of the chaps said to me, don't take any bloody heed of him kid, he was only pulling your leg, so forget it, but I was puzzled over that paint, it is like I said, there was no viciousness in these pranks, but it made their day and another day I was asked to find an old bucket that was used to hold sand in case of a fire, they told me to get the bucket and ask the engine driver could I have a bucket full of compressed air out of the engine. I used to fall for these little tricks but I could see them laughing and after awhile they ran out of these tricks, but they tried once to tell me to go get a glass wedge and I was too wary now to these tricks so they gave up eventually and I honestly believed the first one about the stripe paint, it made their day I suppose, but in fairness they would never put me in any danger, but I suppose it was part of a young lad's initiation into the adult life of the pits, but you live and learn by these good natured pranks by the men, and it shows and highlights the camaraderie that develops among miners to make you feel that you are one of their breed, I am proud to be part of that exclusive club.

Back to Six Bells and to the arrangements that were being made prior to swapping jobs, but the haulage driver was a bit dubious at first of my ability to do his job and how I could handle the engine because of the responsibility of causing a stoppage or cause an accident, which I could understand his reasoning behind his decision, so the opportunity arose one day when the coal was slow coming off the belt because the men I knew by experience that the coal face men were erecting their supports by now, but it enabled me to show him that I wasn't as green as he thought I was, so I sat at the seat and waited for the signal to come, to put a fresh dram under the loading head, the driver sat on

Blue Scars

a bench watching me, little did he know that I was on familiar ground once more and the familiar smell of the grease and oil was like a tonic to me. I will not go into or enter a dialogue of explaining the working of the haulage engine because that would require the knowledge of a mechanic to which on this work I am not conversant with, but on reflection I should have having had so much contact with the moving parts, but I will not dwell on that at this present time, at this moment in time I was more concerned with impressing the regular driver as to the capabilities of working practices, but I omitted to mention my apprenticeship of ten years as I reminisced in my own mind, and when I started to manipulate the controls at the start of a signal from the loading head to haul a dram in a new position he was impressed by my confidence and he relaxed and said to me "hey Chopper you must have done this job before", to which I replied then "I have had the experience at the last colliery I worked but should I have mentioned it you might have thought I was bragging but I also wanted to prove myself that the old instinct was still there, and I was more used to handling a larger haulage in the main transport roads but had the same job as what you are doing now, but I had to prove myself didn't I, that I still had the same skills as then, because a decent time has elapsed since", and that's how we started our relationship together and this went on for quite some time but it would have to come to an end at one time I suppose, when the regular rider had returned and I wondered when the time came to start the job we were signed up to do. Incidentally I did not see much of Dai my in law because we were split up on many occasions, but this did not worry us as lot because we always met at home and discussed the events that was happening during the days that passed, but what I really missed was the pit head baths that I was used to, they were in preparation of being built so we had to travel home in our working clothes and bath at home, this was a little awkward because I was living in apartments at my in-laws home but for a period I was transferred to the night shift for some reason I cannot recollect and at that time it was convenient in a way to live at my wife's parents house which in my case was a little overcrowded. I went to bed after everyone had arisen but I was restless at the inconvenience with one child, but the hospitality of my in-laws was second to none and I suppose you could not expect them to put up with these inconveniences the rest of both our lives, so I looked at the alternative of trying to find a house of our own but I will not go into personal issues, we used to travel on the train from Nantyglo station to work, we all used to meet at the station with other workmen, all in our working attire, I can remember a few of my colleagues I used to travel with, one was Granville Hancock's who lived in the farm house and my next door neighbour Harry Jones, and a few from Brynmawr whose names I cannot remember. We were a mixed assortment of married men and single chaps but the same objective down the pits, this went on for a while in the usual routine getting to pit top, having a puff at a woodbine and hiding our fags and our train tickets in tins of various sizes before beginning the shift, hoping that no one could find our fags in the meantime because the surface men would be on the scrounge should they run out of their own fags and this went on for a short time until I went days and that altered my routine completely. I then had to travel by service bus then which put me in a very awkward position owing to wearing working clothes because when you got onto the bus I had to stand near the door to allow the other passengers to descend and also others to enter, by the time I arrived

Blue Scars

home I thought my one bloody leg was shorter than the other and that went on for a few years until they had completed the pit head baths, to the relief of everyone including myself then I moved to Cwmcelyn but a few months before the opening of the baths while living at Cwmcelyn I still had to bath at home where the absence of a bath at my in-laws house made it more depressing by the inconvenience of the usual old tin bath hanging outside, and the buckets of water ready on the boil. I lived at 66 Cwmcelyn road, a row of houses called Black Rank, for seven years and there was the arrival of my second child Elaine, I used to call our row Mildew Mansions, that stood opposite the feeder ponds, but moved in 1954-5 with my children David and Elaine to the Coed-Cae estate, my address at that time was No. 48 the New Estate, there were no street names at that time because most of the roads were not built and the firm of Cowlins Led was completing the site, the houses then were not the property of the council until completion of most of the site, and roads were formed into various avenues that was sixty-one years ago by 2006, but at the time of handing over to the council the street names began to take place, for my street it became Verwey Road and my number became 12 Verwey Road, and I still reside at this address even today so I was the first tenant to live in this house at the Coed Cae estate to which my rent was 14 Shillings and sixpence per week, 5 shillings and sixpence short by a pound, the black rank houses of Cwmcelyn have long since been demolished to make way for old age pensioners bungalows, but I enjoyed the seven years of good neighbours when I lived there, different neighbours have changed hands and houses at Verwey road in the sixty one years I have lived here, there are only two of the original tenants now left and that includes myself.

How long that's going to last I don't know but hopefully another 50 years. I never chose pit work for my son who has grown up now as with my daughter because there are more opportunities today to choose from where there were none years ago. Back to the pits and I adapted myself to my new environment after leaving Beynons and the coal face and haulage driving and I wondered what was to come next, in the new colliery that I was employed at but I have gone through parts of my work in the previous paragraphs. A new approach would probably be made in the years to come because these changes often occur in pit life as I was about to find out later on, but in the meantime things were just about the same and I had a job and that was most important and there did not seem any progress in what we were signed up for, Dai and I, but we had to be content with the usual chores and you must not be too complacent with even the jobs that you were familiar with because familiarity breeds contempt and that laxness may be your undoing and like the oldsters I used to work with had the saying "keep a cool head and a sharp hatchet" and never relax at any time. It was a good saying that I always abide by, the only time to relax is when you step off the cage and into daylight. So one day when I was waiting for orders outside the officials station when I was approached by a senior overman and asked would I be interested in trying some piece work or contract work if you want to call it in other words, I said what sort of work would that entail, he said it would be pipe fitting to which I had no knowledge whatsoever, he said it would involved face work because one of the team of pipe fitters would be on the point of retiring and it would give advantage also of a higher wage. I thought to myself that it was an opportunity that I should not miss at least I would be working for myself and not be buggered about doing those sundry jobs without the hope of bettering myself and with the incentive of a better pay packet, so he said "think about it Chopper, but don't leave it too long as there were a few interested", so I decided there and then and accepted the chance to try, this was in the middle of the week and at the end of the week he approached me and said "don't come into the station but ask for me", If I can remember his name was a Mr. Arthur Bainton, but anyway he directed me to a small cabin a few yards up from pit bottom and for the pipe fitters, so this I did and was introduced to the team of four men. After I explained who sent me and what for they accepted me because all of us met over the years and knew each other by sight and knew of each other so after a few minutes conversation they told me what was expected of me, and names were handed around and I name the four men now, one was a Mark Small from Six Bells, another was Hayden Lewis from Abertillery if I can remember right he lived in Bridge Street, another Joe Sailes and last was Mr. Bill Barnes from Yellow Rank Cwmcelyn, Blaina, so after the initial get together, I watched the preparations made for the commencement to start their journey into the districts where they were to perform their activities, but I would not be expected to do anything but watch the procedures that the men were about to do, but every man worked individually at their separate tasks and was told to watch my back as they had no time to look after me, because they had to watch their own backs, by the release of maybe a pipe that was holding up debris that was resting on the pipes that had been dislodged from the roof, it needed extra attention. This was a coal face of a length that exceeded the coal face that I had been used to in Beynons. This face was approx hundred and twenty yards long, this is why I have to give graphic details of work performed by some face workers, including myself, otherwise there would not be any sense in me explaining the dangers associated with men in confined spaces and men trying to complete contracts that allow the coal face to move forward and should at any time the coal face be held up for any length of time then mother nature would certainly put her foot down and the pressure of the supports would not be adequate to prevent the extraction of the pipes and conveyors more difficult, so it was in everyone's interest to keep the face moving as to avoid any such incidents that may present itself. So with these thoughts in mind I intended to take special notice of what my mates were doing in which it was no easy ask to concentrate in the dust that was left by the sootiness of the belt conveyor after the structures had been moved to a new line in readiness for the cutting machine to cut the face, the boys were carrying lengths of wire with them, I wondered about that, but the reason was to hang the dismantled pipes in the new track above the conveyor, another task that required the threading of dismantled pipes through the supports and some of these pipes were eighteen feet in length, I followed my new mates up the face observing every moment but helped in the threading of pipes into the new track, it was some consolation in trying to keep in which it was appreciated, by comments of "good lad", it was all I could do anyway, I did not have any spanners to unscrew the clamps that tied the column of pipes together it was not my job and I on my first day was only there to observe the procedures and to gain the knowledge, when one of my mates retired and I was to take his place in whatever job I was entrusted to do, it was harder to learn than doing the job yourself at least you knew what your task was. I was not paid out of their contract of four men but I was more or less paid the equivalent or near enough to their level because I was in the team ready to take over when needed so it was up to me to gain that experience but after completion of the day's work we went back to the fitters cabin and the preparation of cutting the wire lengths needed for next day's excitement. I was asked "what do you think of it Chop?", I said "I'll get over it I suppose it was a bit strange" I said "but I will have to get my act together to compete with you bloody lot" and they started to laugh "bit different Chopper from engine driving chap". I could not agree more but I carried on until the day came when one of the men came to his final day and he retired and my day of being carried about by the team came also to an end, and I was initiated to become part of a regular team and to take over the role of our departed friend, but just before he left he said to me "I have no use for these spanners Chopper so you can make better use of them than I can, and it will save you buying a set and the best of luck". I thanked him for thinking about me and the last words he said to me was "don't get intimidated by one of the team who may try and use his sort of bullying tactics on you, because he has tried it on all of us in turn and you being a new hand at the game, he may try it on you, that's my advice to you Chopper". I told him I would bear that in mind but I told him that I had a fair idea to whom he was referring to and soon after that he left us. In one way I was sorry to see him go because he was a decent old stick and so I became one of the gang on the afternoon shift and waited for our next assignment to one of the districts, but the words of wisdom still rankled in my mind from that decent old stick but as always out of groups of men someone always assumes the role of leadership either by persuasion or a bullish attitude in which you will in any industry get. I was hoping that it did not apply in our case but we will see in due course I expect, not that it worried me to that extent, if anyone wants to take over the responsibility that's alright by me, all I was interested in was the task ahead but it was a nice feeling to walk by the officials station and think that you were your own boss in a way and not to worry about orders, to direct you to different jobs, and walk to your little cabin and wait till you were needed, but on some occasions we were not needed then we would prepare for the shifts that followed but you certainly had to make up for it when two faces needed your attention, but that did not happen very often. So the time came that we were needed and we lost no time in getting there to extract the pipes, the quicker you got them out the quicker you were out of danger, that was the first priority so off we go, on arriving we stripped off our coats and entered the face which was already in chaos by the conveyor shifters rolling up the belts and getting the conveyor structures in a new line, they were half way up the face and that gave us the chance then to uncouple the sections of pipes. My job for the time being was to go on up the face and uncouple as many as I could in the wake of the conveyor shifters while the rest released the pipes and threaded them through the supports into a new line, but as I proceeded to uncouple my eyes were like a bloody bird, watching the top over my head and in the job (waste ground left behind by the absence of the conveyor) but also recognising the taunt wires that signifies that there was something heavy resting on the pipes over my head and by releasing the clamps holding the pipes together it may also release some debris held up by those taunt wires, and there was not a lot of room to manoeuvre in that low area, so I was ever wary of those little traps, I tell you my guts was turning over every time a stone fell on some of the conveyor structure nearby. After I had uncoupled a voice shouted up the open end of the pipe "come on down Chop and give a hand", the voice coming up the empty column of pipes sounded as if it was coming from the darkness of the waste and just could not make out where it was coming from until it dawned on me the source, so I tucked my spanners in my belt and crawled back down the face to the boys and was grateful for their company, so two of us helped to thread the pipes into the new line while the other two did the same until the place was reached where I left off. I explained about those taunt wires holding the last two pipes and they said get behind while we drop the one end of the pipe in which I had described the weight, as soon as they dropped the one end, bloody hell down it came, I thought the whole place was falling in on us because the pipe was holding back a large stone which was followed by a cascade of rubble released by the falling of that stone, and I thought I was kneeling under that pipe a few moments ago and what if that bloody wire had snapped then. I could not bear to think about it and of my lucky escape but it proved that you have to keep your wits about you at all times. Anyway after completing our job we had five minutes whiff and made our way back to pit bottom where we had a bit of grub and I examined my arms and hands, they were scratched everywhere and my shirt on my back was ripped in a few places where it had caught by trailing wire ends that could not be released owing to being held fast in the pressure on top of the roof supports, and that was my first introduction to pipe fitting at face level and this is why I am giving my true account of the work of a pipe fitter and to give you the atmosphere and the problems that can cause a lot of these prolific nature of accidents that occur and often happens when a less competent person like myself are first introduced into unfamiliar jobs, but you learn pretty quickly when confronted with your decisions and carry on with the little knowledge that you have, but sometimes under the supervision of a more competent person and like the favourite saying of politicians, lessons can be learned from listening but in mining you have to conform by the rules that are there to protect you, but you also have to abide by your own initiatives and when you perform as a face pipe fitter, when you uncouple pipes, you place the clamps in such a place so that you have no need to hunt for them after, thus avoiding putting yourself in a second danger of perhaps the danger of shifting strata of the place that you vacated, therefore you should place these clamps where they are visible and not buried by other workmen working nearby and in turn could impede the progress in completing the contract on time, so everything had to be placed in their respective places for when the time came to assemble the pipe column together, but I learned these small things on my way to be a successful pipe fitter and apart from a few shouting matches from my butties saying "come on for Christ sake or we'll be here all bloody day", but most of the comments were of a friendly banter and not of an antagonistic conflict out of frustration, but my lack of foresight on my behalf to certain changes that present itself at a particular time, but all in all everything went fine after a few pit falls you acquire whilst learning your apprenticeship. Every yard you gain on piecework made that extra penny or half penny and in old money every penny counted. This face we were turning was approximately two hundred yards long and prices were paid on what the circumference of the pipe you were turning, this is an example, for the turning of one inch pipes a half penny a yard, two inch were a penny a yard, 3 inch were one and a half pence, and most contractors were paid on percentage rates and most pipe and turnover conveyors fell into that category. All I know is our percentages were on a 84% and other 16% percent were added, so as far as I was concerned was the yardage we earned was doubled, but I knew little enough at that time anyway, all I know is that our count for the week was entered to the time keeper at the end of the following week, it was put onto a master ticket that showed the wages for the week, and we all had our separate tickets to show equal shares but I won't go into that as long as my wages corresponded with the rest, I wasn't bothered.

The only thing that did make me nervous about the work was when two faces wanted turning on the same day or afternoon, that meant we had to do a double shift that not only would we come into contact with turnover gangs and the cutting machine had cut the face for the following morning and the crashers (the men that retrieved support for the job waste) to be used again at the face, would obviously be disturbing the entire face by retrieving those supports from the waste, therefore there would be two gangs turning over and the gang would be crashing the waste and we would all be working in the same vicinity.

Chopper Davies

And I did not relish the thought of what we were going to meet when we entered that face, between the already cutting of the face and the start the crashers had in the time we had our break between shifts at the canteen to replenish our food supplies and I anticipated we were in for a rough time of it but that depended on reception that met our eyes when we entered the face and all I knew that I would be glad when this bloody night was over, but my fear proved groundless because the crashers finding our pipes impeded their efforts and was detrimental to their own safety had dropped a few lengths of the pipes and pushed them against the new line of props (supports), so our intentions were to uncouple these lengths of pipes and slung them into the new track but not tying them yet, but undid another twenty yards or so and precisely the same thing of slinging them in the new lines and matched them up with the ones that were already there, while my two mates continued to dismantle another few yards we other two started to couple the ones that were already in the new track and made a getaway from the crashing party, but aware of the cut coal that were showing signs of breaking up in parts that could then dislodge slips of coal and therefore they in turn would dislodge properly supports that would bring down the pipes that were hung on the wires but fortunately in our case did not happen but the supports in this face were steels which supported the steel flats above it, so the newly erected steel supports had not been subjected to the new weight as yet so was easily dislodged, this problem also faced the conveyor shifters but they had advanced considerably to get out of this sort of danger which was activated by the release of supports by crashers. Their job I did not relish but everybody was in the presence of the same danger, but we had another hundred or more yards to go yet but we eventually got over that owing to the co-operation of some of the conveyor shifters who had almost completed their own task but the poor old crashers were quite a way behind, thank god for that, not wishing them no ill will but it put my nerves in more of a stable state that I started in, the only finishing touches we had to do was to put the ends on the open ends that meant traversing the face once more, that was scary enough but the job was done and I didn't care a bugger. After that we sat out in the roadway to have a few minutes whiff, the little roadway was only about five feet in height but to me it felt like I was sitting on the bottom of the pit but it was an airway road that ventilated the face and the district but in that quarter of an hour your thoughts began to wander and your eyes started to view your surroundings, thoughts would start in your head while sat there in that little haven was that all this was held up by these wooden timbers that held up mountains, roads and buildings and millions of tons of earth and we sat there like bloody cave dwellers, it was not a very good place to have thoughts like that and specially for anyone with claustrophobic minds because they would go bonkers and that would never do because if most miners suffered that way and attitude then there would not be many miners to work the pits. While I sat there, there was a flash of quicksilver in front of my eyes, it was a small grey mouse covered in stone dust from the side of the roadway and darted into the many crevices of the stonework that helped to bind timber and stone together to resist the pressure of weight above our heads and make it more resistant. The mouse was a small part of the denizens that made up the world of inhabitants that fester the pits, including cockroaches that were not of the domestic type that is in our homes and deny any attempt to squash it and there was an assortment of flying objects that get into your hair and found their way into your clothing that if not noticed could be carried into your homes, so you had to be careful and search your clothes if they were left at any time for a long period. They came out of the decaying timber that may lay about the roadways, you can expect that when these creatures survive in some of the damp roadways of the pit when all at once I was shaken out of my reverie by a shout from one of the boys 'Come on Chopper or we will bloody leave you behind', I thought that will be the day but if they had left me which they wouldn't have of course I would have no idea of getting out of this bloody hole. So off we wended out way out through different short cuts that I never knew about, in this labyrinth of small tunnels that were only known to ones that were used to the run of things I know I would have been lost so we were on our way to pit bottom which was two miles away by my reckoning, we were dodging broken down timbers overhead which had succumbed to the pressure above but were able to hold some sizeable chunks of stone but could be dislodged at any moment but I think the ground was more settled by now because we were a mile or more from the actual workings so there was small risk of too much pressure being felt this far back but who knows what mother nature feels like being disturbed at this hour of the morning, but these little tunnels were only used as airways as a temporary source of air, I suppose but I don't know these things so I am not able to comment on it but I only know it was only used by the likes and whoever travelled that way except for an official doing the airway rounds on his examination and checked the tunnel in case there was a fall of ground that may block the air, but it makes you wonder what sort of men, young or old, passed through these same tunnels over the years when the original tunnel was formed, it makes you think what thoughts were in their minds at that time but they perhaps are gone now. After a few more twists and turns and came in sight of our goal pit bottom and our little cabin and made preparations for our next we cut lengths of wire and renewed a few fresh bolts to replace some damaged ones that needed replacing, there were different types of clamps that were used in connecting the pipe column together such as Carlton, Sutton Williams and a mixture of universal joints, there were different size bolts used for each clamp and type of reference determined the type of spanner to be used but came automatic to the regular users of these tools but we had to use too, namely Colliers tools, Mandrels (picks) shovels to release buried points that were hidden by falls of debris but always returned the tools to whatever place we had borrowed them from, to put them on their respective tool bars and unknown to the colliers that we were borrowing them from, some of these tools were always securely locked on bars by padlocks and methods were used to open these locks by manipulation of pieces of wire which I found out the men were a crack hand at in acquiring the tools but the use of this method were frowned upon by management and also the union (committee) by the unauthorised use of someone else's tools of trade because like I stated earlier, the hatchet was the most dominant factor of the colliers trade but in steel supports the main tool is obvious, that the tool used then was the sledgehammer that erected and hammered the steel supports into place so the hatchet took second place then, only to cut lengths of wood to place between the steel upright and the steel flat above it to prevent the friction of steel against steel should the upright get discharged, but anyway I was surprised at one time by the attitude of one of our gang who had no scruples in leaving these tools around without securing them in their proper place on the toolbar which reflected on all of us and illuminating the fact that the tools were used by us if we were the only ones that were in that vicinity at that time and his remarks by saying 'OK bugger them' to my view was not very good practice and drew attention directly towards us who were relying on the good nature of some of the colliers and did a lot of animosity towards the lot of us, so I voiced my concern to the other chaps who also did not condone these actions and when approached on the subject the chap in question gave a negative reaction and used language to that effect and I remember the conversation I had with the chap I took over from about the one bad apple and I realised the one that had come to light so I bit my tongue being new to the gang not to cause any friction and not only that, but as a senior hand might use his influence to get rid of me so I was in catch 22 but when I was firmly established into the gang I would certainly voice my opinion and they would find out that I have a mind of my own, as one of the chaps said to me 'You have got a bit of a temper haven't you Chopper', I replied that I could stand on my own two feet if necessary but I haven't had call to use it among a good bunch of chaps and the man in question was half my age and told him in so many words and told him on one occasion not to push it too far, but in one sense I wish I hadn't opened my mouth and in working together in a confined and dangerous low place you did not need any animosity but relied on each other's responsibility to each other, but he had other means of getting back at me, subtle little things like uncoupling the wrong side of a joint and leaving me with the weight of the three inch brass cock that was used by the cutter man, couple his lines or bags some prefer to them, they were and resembled three large tubes not unlike your hoover bags on the vacuum cleaner, in fact identical only on a much larger

Blue Scars

scale and these brass cocks were cumbersome and awkward to manoeuvre in cramped conditions. I did not complain because I was a lot younger than him that's why I relented a little but I think my other two butties were a little afraid of him but never showed it but were quite aware of his pushy attitude in a manner of speaking but like I said there was always that one to upset a team and I believe he was that one with an attitude problem but it comes to a point where you can only stand so much and to the amusement of my two other butties confronted on one occasion when he needed me to a point that something had to be done and asked who do you think you are talking to and that I wasn't a kid of eight, he looked at me and realized that he was getting too far with his badgering and didn't care a bugger about him and I was half his age and I had come to my boiling point, he showed his true colours and sort of backed down, not that he was frightened of me but he knew that he could not browbeat me into his way of thinking but we got over that and never pursued the argument anymore and that little confrontation soon evaporated out of our system but is shows that you must not knuckle down to some people.

But to bring back memories of good old Beymore Colliery and some of the tricks played on some unsuspecting men but I was not included in some of these so called pranks which in some cases I found distasteful especially to some of the recipients and did not work out very favourably in some respects but not in the way of vindictiveness nor nastiness but to what taste the prankster had on his mind and resulted sometimes in heated arguments that I witnessed as a boy and when the tricks were not very well appreciated in which I did not blame them because I would have felt the same way, such as putting objects into the sleeves of coats and when it came to walking home to pit bottom but try dislodging a hay or brick or a lump of wood from your sleeve when you are in a hurry to get out it would be more than an annoyance I can tell you and that was one of the tricks that did not go down very well, imagine the frustration in someone trying to dislodge the offending article from the sleeve in a rush to leave but the culprit was never found out in fear of the consequences of a man walking out with one arm in the sleeve and the other wrapped around him or another trick if you can call it that of when there was an expanse of water that needed waders (wellingtons) to cross it, men used to take off their boots and use waders to cross the water and by the end of their shift crossed over to don their boots to find that someone had nailed a boot to the wooden sleeper of rail track, I thought as a practical joke that joke was taken out of context and a pitiful attempt at humour and didn't need recognition as a prank and certainly not appreciated by the recipient of it but these things do happen and again the anger and frustration that resulted from it, but these jokes were few and far between and again the culprit was never found and that was some of the so called jokes that I had heard of but nothing to be proud of.

At Six Bells after my confrontation with one of the team, one of the chaps came up to me and said 'You sure shut his bloody mouth up Chopper' to which I replied 'Forget it and don't open up old wounds and let's work together amicably as a team' and so we did not bear a grudge and funny enough turned out to be one of my best mates and I looked to him for advice, him having the experience and he could see that I would go out of my way to help also if he was in any difficulty but he did not swing the heavy ends on me but took it in turns to when and where that occurred but

Blue Scars

it proved one thing that sometimes you have to stand up for yourself, not that it was needed very often.

But if there were not any turnovers we were all split up to do various jobs, some were put in cleaning conveyor belts that of the main carrier road that spillage had occurred and others would be sent on different sundry jobs that I was quite familiar with, this was to supplement our income and paid at the rate of a labourer. Sometimes I would be sent repairing broken rail tracks in which I would be alone for hours and sometimes things get a little eerie being on your own, the only light that was there came from your own lamp that threw distorted images and shadows from the darkened recesses of manholes and different objects projecting at different angles and I would jump a foot when a dislodged stone fell on the casing of the hollow pipe. Sometimes your eyes played tricks and your imagination runs riot but sometimes the silence was broken by the timing of the deputy doing his rounds and having a five minutes whiff and a final check of the work I had been instructed to do but I enjoyed these breaks in between the turnovers instead of being keyed up in the face where you had to be like a bird in the gardens your eyes had to swivel from side to side, there is no other way to put it but the object was the same in both, the bird and myself was to get out of the danger zone. My next job was a bit more complicated in between turnovers it involved being or trying to be a haulier for the day by taking the pony to a place when it was needed to haul trams of debris from a place that was not accessible by rope because of the twisting and turning of small passageways and that was the only way of getting supplies to that area and this job I had never undertaken in all the jobs I had previously done in the jobs I had been put to it was one of these jobs that I was ready to baulk at, not having dealings with horses before and I am not afraid to say that I was scared of horses because they were unpredictable, not like the handling of men or tools, some of these buggers had minds of their own and were used to the command of the regular hauliers and they knew a stranger and they could sense whether your attitude was of command or fear in which my case were the latter and I could not disguise it and horses could sense that, or so I was told you had to be firm in your approach so I entered stable and view around me and looked at the few horses in their stalls and decided I would pick the smallest that was there, but afterwards I was told 'You picked the wrong bugger chop, he is one of the craftiest of them all'. I thought why didn't they inform me before, anyway I eyed him up and put my most authoritive voice I could muster, move over, not a bloody movement, I knew I had to go aside of him and squeeze by his side to undo the rope that was tying him to the manger but every time I tried to enter he would my side all the time and the clicking of them shod hooves deterred me for a while but time was running out and I wasn't getting any results so I made a bold bid and forced him out of the way and managed to untie the rope holding to the manger and backed him out of the stall. I thought to myself good boy Chop you have got him, my next job was to harness him then I thought of those cowboys and how they used to sling their saddles over the backs. That completed I tried to put his collar on, the cowboys never had to do that but there was a knack of doing this I found out after of course but not at the time I needed that advice, so I put the collar over his head but it got stuck half way. I thought I would be choking him so I left it alone till later, so far so good, so I led him out of the stable and shut the door in case other horses may escape and while doing that I looked around and the horse was on its bloody side, they told me later on that if the house saw a bit of dust and hay mixed up outside the horses would rollover in a dust bath, get up after a while, shake themselves and wait for further commands, this the horse was doing but on the ground was left the harness because I was supposed to have tied the so called girth strap under his belly. The horse waited and I am sure the bugger was laughing at me so I tried again to no avail and I did not even attempt to put the collar on him again. I didn't know how far we had to go till I met someone farther on, so the horse trotted on ahead of me and I was dragging the damned harness and collar behind him but the horse knew where he was going being used to the direction. I followed him with the tack, the horse arrived there before me and when the blokes saw me in a bath of sweat and carrying the collar and dragging the tack I never heard so much laughter coming from me, but I wasn't laughing, but fairplay they did all the work of fitting up the harness and I watched the deft way they twisted the collar on its neck, after that they attached the contraption around the horse and to the dram. They called the contraption the shaft and gun and after they completed what work the horse was sent to do, dismantled the lot but left the harness and collar on him and started to lead him back to the stable but one of the fellas said 'Leave him to go Chopper he will find his own way back and I marvelled how the horse could find his way back in the inky darkness to the stables but he did because as soon as I tied the reins to his collar he was off and by the time I had reached the stable the horse was stood outside and when I came up to him the door opened and a haulier put him in the stall after unharnessing and put it on the hook that was available then throwing the collar on top and said 'I'll see to the rest'. I couldn't thank him enough, but all he said to me was 'Bugger off Chop and catch the cage', to which I was not long in doing. Thank God that ended my day as a haulier and wishing that we had a turnover tomorrow but I suppose you have to take things as they come but it's all the experience of miners work, well mine at any rate and if the pits were still open and I was young enough I would be one of the first to go back there but I know there is no hope of that Maggie Thatcher saw to that but I still maintain my coal fire, she hasn't stopped that which gives me warmth and homely atmosphere that can never be replaced by gas.

The only time that I ever went back to the stables was at the time I was doing a job at pit bottom during in which time I was frozen not used to be exposed to those elements and grub time I went to the stables to the warmth I knew to be there. I smelt the manure and ammonia that emanated from the horses, it was an agreeable smell and as I entered there were the usual array of harnesses on their hooks and was met by the hostler that looked after the wellbeing of the horses and to see if they were in good order. Well on seeing me I was given a look saying what are you doing in here, I have only come in for five minutes to get thawed out, 'This is not a library' he said 'and I didn't think that you wanted to come here after what I have been hearing about you, it was the biggest laugh that they had ever had and no wonder the horses kept sniggering but they have their code of their own I expect but at least you did no harm only to yourself I suppose but if you keep coming in here others will be expected to do the same so make it your last visit do you understand or I will have to report you so go and have a sit on that box for now'. 'Well thanks' I said 'I will try to remember' and I was glad in a way because he would not let me handle any more horses, that put my mind at rest so after

Blue Scars

about 10 minutes I went back to pit bottom to finish my job, it was bitterly cold, it would have been better not to go to the stables because I felt the cold more so now. I should have gone up to our own cabin a few yards away, how the hell these blokes stuck it I don't know but I suppose you could get used to anything in time but I was sure that that sort of career would not appeal to me, so one day while we were cutting wire in preparation for our next turnover the senior overman came into the cabin and said to me 'They tell me that you are a fair hand at haulage work Chopper' well I said 'No better than anyone else'. Well he said 'I've heard different from a good source so I've got a proposition to put to you that when there are not any turnovers I would like to give you the chance of learning to drive electrical haulages and you already have the knowledge to do so by all accounts, so how about it'? Well I thought it better than being buggered about so I accepted, he said 'You will only be needed then that if an engine driver was absent or ill, you could fill that post but only after your turnover was completed, so in between them you can learn'. I said fair enough and arrangements were made to that effect, I did not tell him that I already knew about electrical haulages from the past but declined in case he may think I was bigheaded but what was different was to control the roads, that's what engine driving is all about, mastering the roads. The driving then becomes automatic more or less and so it was settled that the first chance I get I would be available, it would be a better number than buggered about everywhere and I was happy with the arrangements and waited for the opportunity, but some electrical haulages had different controls, some with levers and some with wheel control but that in this case only applied to the largest of haulages and I only had to graduate on the smaller ones till I reached that stage and that would come in the process of learning. So sometimes the pipe fitting finished earlier some days because of maybe a bit of luck or maybe we had all got used to our job, I don't know but sometimes it did happen and I had time to visit some of these haulages working which gave me a better advantage by viewing the procedures but had to be careful because you were not permitted to travel on unauthorised transport planes, only if you had permission of which I never had, at least not at this stage, but it would be granted to me by travelling from one engine to another, so one day the deputy approached me and said that I was to report to the 75 hp at W porting and gain some knowledge there, so I did that but on entering the engine house I was shouted at by the driver 'Can't you see the notice that no unauthorised person was to enter', he didn't give me a chance to answer because a signal came and he had to use his engine, he waved his hand to tell me to get out so out I went and strolled down the roadway, keeping my eyes on the rope ways and when I could see in the distance the coming into view I quickly went into a manhole at the side of the road, after it had passed I ventured further to have a view of the undulations in the rail track which would give me some idea of how to control the engine when it was time for me to try it myself when the time came, so when the next journey of coal passed me I followed it out to near enough the engine but I was greeted by the driver a bit different now, who in my absence had received a phone call to ask had I arrived there. I entered the engine house but with a different tone of voice by the driver who did not know that I had been sent there to observe. I was glad to get in there out of the cold wind which howled through a narrow part of the roadway where the engine was situated. 'I'm sorry about that mate' the driver said 'Because you have got every Tom, Dick and Harry wanting to come inside to have a warm'. All I said to him was 'I don't bloody well blame them'. He said 'You have got to stop that in the beginning'. I understood his reasoning that he can get distracted by them when driving and I thought again I never got distracted when I was driving, I was always glad of a bit of company but I suppose everyone had their own different views so I settled down and observed his actions, but I think he was trying to show off a bit trying to impress me with his skills, he was darting about from one side then another but I was more concerned with the undulation in some parts of the roadway that I would have to try and remember, but an hour or so later when I was more or less confident that I could handle this but I would have to feel the experience myself by the handling of the engine because of the different surging of power of compressed air and electrical surging so I watched his performance when handling a derailment and I was quite impressed and told him so, to which he accepted the compliment, we began our relationship by questioning each other of what sorts of jobs we had been given during our employment, his answers were mostly about driving and I told him of my experiences of work including the basics of driving at the previous colliery and to the jobs I was currently doing as a pipe fitter but was asked to learn to be a spare hand at most of the haulages and he said at least you will have a variety of jobs whereas I am bloody stuck here and all I said to him was at least you have a permanent job and you to be thankful for that and the men at the other end of the rope can rely on you for their safety and we all give a hundred percent in doing our jobs, that's the main priority.

I was given the chance on one occasion when the journeys had slackened down to test the knowledge I had picked up at 75hp haulage and proved after a few more runs of the roads was more or less confident that I would be able to take over when the occasion arose and that was my start in a selection of jobs but wondered when Dai and I were going to start on the main job we were signed on to do?

But afternoon shifts were different than day shifts to me anyway because when I got out of bed early mornings I needed an hour or two to come to myself and if I didn't have that time it depended on what mood you started the day and I needed that couple of hours to have a cup of tea and a few woodbines to come to myself, so as soon as you entered that cage to go down the pit the thoughts of a fag never entered your mind but other men found another way to satisfy their craving for a fag by the substitute of chewing tobacco. That was the nearest thing to a fag yet it served its purpose. I also tried this method but at the first chew I must have swallowed a grain and resulted in being sick as a dog or whatever the dog felt like and that put me off for quite some time I can assure you, but eventually succumb to the temptation of the filthy habit but accepted it a bit more cautiously now but finally became addicted to it the same as other men, but when the shift was over I was the first to rush for my fag that was in my hidden tin, hoping that no one had found my hiding place, the searchers of contraband did not have a chance of finding any on me because my fag was hidden away at the bottom of a girder near top, anywhere that you could find to conceal it because if found on your person then it would be you for the manager either for a reprimand or the sack depending on how many times you committed the offence, but it would be a fool to take such a risk, not because it was the cigarette but the means of lighting it at the risk of other men's lives, but men have different personalities and should be screened before employed in deep mines but today's standards they would be classed as racial or attacks on their human rights but fair play these instances rarely occur and during my years in the pits never encountered such infringement of these rules.

But let's go back to pit bottom, these men were the first to start and last to finish and I was glad I was not picked for any of these jobs, I don't think I could have stuck it, especially near the bottom of a downcast shaft where the air was drawn down in the effort to circulate the whole pit, it was cold in the summer without the coming of winter months, they must be as hard as bloody iron and in a slack period showed an expanse of the rail track and I marvelled at the way the track was assembled and how the diamond patterned track were interlocked into each other and in such a way that individual drams were separated and directed to their respective cages to ascend the shaft and then returned to be reassembled in their long line of empties that made up a journey and sent back into the labyrinth of tunnels into the bowels of the earth time and time again in that continuous flow of coal from distant coal faces, no wonder the men on the pit bottom worked like robots, uncoupling (shackling) and coupling the drams and switching them to their respective cages and received and entrusted to the onsetter (hitcher) to raise them up the shaft but he was placed in a precarious position while standing between two lines of drams with barely a yard to move about, the onsetters job was to do all signalling connecting the dispatched drams of coal but kept a wary eye on the drams at his feet in case one jumped the track close to his feet but when the signal came from above that it was grub time then the scramble to their cabin to choose the warmest place from the doorway which was covered by brattice sheets, a sacking covered in tar or some substance that kept out the biting draft. And now I could see the responsibility of the engine driver delivering his journey of full drams into that medley of bottom workers, if too fast the contact with stationary drams would be a catastrophe especially if one of the men were stooping between the drams, uncoupling them or to something of that nature (I am trying to deploy a degree of detail into the work as to give you an exact account of the trials confronted by men of the bottom).

Perhaps one of these days it may become one of my responsibilities when I would be the engine driver to avoid the excessive bumping of stationary drams because transport was the cause of prolific accidents by not having enough care in some cases of conforming to rules governing transport but to avoid some of such happenings, safety meetings were held but were mostly attended by safety officials and the union, therefore the impact were lost on the grass roots of the men, by those verbal talks and were not so illuminating to make that impact but could only have the attention of men by visual means, by pictures of paintings that could throw light on these safety values, but that did not come till later, until I was approached by management and coal board by emphasising and made more prominent need for them to take interest in what job they were employed at, my role in this was to provide these small but effective ways of dealing with the problems of safety values by illustrating most aspects of mining life in picture form, this form of visual art was not only to capture their attention but also provided a form of entertainment that were shown at different venues, mostly clubs and pubs, in which miners wives also attended which gave them an insight into what their families were employed in and the tasks they were at, this also gave a powerful impact to the audience to what the dangers were associated with in getting out of those black diamond's that have caused so much heartache to many that lives and work in a mining community, so it was my job really to provide this scenario for such events to take place and with the usual criticism by non-miners who had not any idea what these images represented, but only the novelty of the entertainment but very few left when these little plays were in progress. Some of these plays were played by local lads who took some of the parts they played seriously enough to take part in other safety sketches that developed later on as it gained momentum. As these plays also developed into competitions between various pits as the sketches progressed along the lines of safety only and gave me the encouragement to make backdrops for these budding actors to perform under. It made not better thrill to them that it gave me I may tell you, these paintings were eight feet by eight feet murals that impressed many of the opposing teams but were not real substitutes for real activities, not even museum can capture the actual work on their tourist guide, it only gives you the atmosphere of the pits, I suppose that's what it's really there for, dressed up for its purpose, the guides are quite capable of doing their job but are curtailed by regulations governing tourists and to ensure their safety, but I am pleased that mining museums keep our heritage alive and preserve it for future generations who have little knowledge of a mining industry but my work as a mining artist try to immortalise my friends of all creed and culture in mining communities, at least that is my greatest ambition after spending thirty five years and approximately twenty to thirty hours a week of painting deserves some recognition of my dedication to a now vanished era.

In transport roads visibility is limited to the light of your headlamp and cannot foresee the hidden traps that lay ahead such as a haulage rope getting snagged on some protruding objects such as the rope snagging on a sleeper or an exposed bold in the track and should a traveller on that road be in that vicinity I dread to think of the consequences, the man in question I am referring to could be a track maintenance employee, authorised to travel transport plane in his duty to but as he approached this hidden trap the tension on that rope would increase to such an extent that should it be released it would wipe out everything that was moveable unless the track maintainer could notice the out of alignment of the rope which on a straight length of road would be centralised and not at an angle, these would be noticed by a regular traveller but not to a novice, that is why a notice displaying unauthorised persons are not permitted to travel the road when working journeys are in progress, but this is to illustrate what transport dangers are about and that's only one of them, not to mention of men having a ride out by standing on shackles between the drams, what if that same dram got derailed and threw the leading dram in front of you off the tracks also, it would shake you off the shackle underneath the wheels and also men travelled in an empty drams resulting in practically the same predicament. I know that because of the incident in Beynon's colliery the time that the empty dram turned on its side at a curve in the roadway, these are some of transport accidents, not to mention many more but these are some of the few but I would not change my lifestyle now because mining is running in my blood and friendships that have been created over the years by men from all walks of life but I still remember most of what has happened in the past.

Blue Scars

One day while working at my present job pipe fitting I was called to the phone and my first thought was that something had happened at home because it was unusual to be summoned out of the face while working, it was the deputy of the district who answered the phone and requested me to return to pit bottom as there was a man coming to take my place to relieve me as the electric haulage driver had been taken ill. I very nearly told him to bugger off after frightening me but I was told to report at pit bottom but I don't think my mates will be pleased about this because the man to replace me would not have the same enthusiasm not being one of the team, plus having no knowledge of the work involved, but orders and orders and not only that but that was that they had paid me for in the absence of turnovers so I could not very well argue about it so I had no other alternative but obey orders and abide by the request, but it was not the same haulage I had been attending while I had been having a rest from the turnover so I wondered if it was the same haulage, that was all right but was it, I was soon to find out but luck was with me and I did the driving to everyone's satisfaction and I remained there until the return of the regular drives but in the mean time things altered to the effect that after turnover now I was put on other jobs, as with the others, but with complete contrast to driving my next job was to help the timber man in the airways and that like I said was in complete contrast, it involved a bit of heavy lifting and it was the heaviest job I had undertaken so far. It was repairing broken timbers and replacing them with new timbers, the timbers that were being used was in lengths of about eight feet in length and required lifting on to two uprights of the same length but it also needed the assistance of other timber men who were working perhaps about fifty yards away, that's how heavy some of these timbers were and I knew by the end of that shift and for days after I would feel the effect of that couple of shifts and I was right, my poor old muscles were tied up in knots and the backs of my legs from straining upwards and stretching my back was in half having not being used to such heavy lifting and no matter how technology has changed it was no match for mother nature when she puts her foot down, by the look of some of the carnage that the roadway was subjected to by the splitting of those massive timbers by the weight they had been subjected to and again I wondered of the weight that lay above us, but I suppose the exploitation of this earth had to be done to expose the hidden wealth of coal reserves to keep the country supplied and depended on, not only that but the bye products the coal produced and which nature had provided but at the cost of so many lives to produce this wealth until we have an alternative power resources but our blue scars will never fade away and will always be with us, of men who have spent their lives at the coal face, a legacy that will take a long time to erase and no amount of soap and water can wash away those blue scars, the badge of the miners.

Well the world have changed a lot since I was a kid and what I can see of it not for the better either, you have climate changes, seasons change but most noticeable changes are the people, they say you cannot live in the past but to look forward, forward to what? We have a multi cultural society, the age of the motor car which contribute most to greenhouse gases where in the past were mostly horse drawn vehicles, of which we don't want to go back to that because of economical reasons but myself in my day never envisaged the diseases that are rampant today, for instance MRSA, e-coli, not to mention quite a few, so you cannot

Blue Scars

blame coal but chemical induced products, too many to name and I defend the right to defend coal and say carbon emissions are not entirely to blame in coal but contribute to society in the by-products that are derived from it but I will not concern myself at present but try and concentrate but relate to matters that do concern me to try and illuminate the age when coal was kind and my defence of it. They try to find alternatives for energy when there are huge reserves of coal available and could be made smokeless with the technology that is available today.

But today old habits have changed from the time people and games of the past are lost to us, when I was a kid games like 'bat and cattie' where you placed a length of stick into a hole in the ground leaving the stick over the lip of one hole and strike it with a length of broom handle while in flight ran to boundaries that were placed in a circle, then back to the delivery end again, similar game to baseball that was some of our kids games. We had games of rugger and football then but our ball was a rolled up Argos or the football echo tied up with string for footballs were not easy to get in them days but if they were there was not enough money around to buy them anyway and our goalposts were a coat off your back or a jersey, anything that represented a goal post, that were some of the games, there was not end to kids entertainment but all these forms of entertainment are now distant memories and times have moved on now to computers, laptops and play stations, the microchip age where children of today never knew the happiness of a bit of dirt to play in but flowers grows out of dirt and it certainly did no harm to my generation but we did have one disappointment, on one occasion, in the early days, those days cigarettes were sold with photos inside of photos of every kind of memorabilia, of such things as kings and queens, cricketers, ships and a selection of everyday sporting personalities and in shops that sold sweets they had slabs of nougat about four inches long. I can remember the pink and white slabs of nougat but it was the wrappings in which they came in that attracted us most because of every wrapper were printed the names of football teams to which you had to collect for a prize, if I can remember rightly you had to collect twenty teams names, denoting the towns that they played for, our halfpennies and pennies were spent on these wrappers because the prize was a football, everyone was swapping these wrappers trying to get the whole set of photos but the most elusive was Portsmouth. I'm damned if we could obtain that one but after a bit of bribing and swapping finally acquired a full set so my father sent away for this prize and the gang awaited the arrival of our prize and to our delight it arrived after a few weeks in a flat cardboard box, we could not understand how a football could be flat but on opening the box found the casing and bladder were pressed flat, it looked perfect in the brownish yellow and we could not wait to get the ball inflated but with the use of the bicycle pump that problem was solve and did not regret the money we had spent, we discarded our paper wrapped football and made use of the school playing fields, sometimes we played in the street to the annoyance of some neighbours when the ball accidently bounced into gardens and into the flowers but there were not many places that you could give the ball a good belt, this went on for weeks until disaster struck while kicking the ball one day it bounced off a wall and careered down Garn Road and the boys ran like hell to catch up with it, some falling down in the attempt to stop it but to no avail the ball gathered speed down the steep road defying attempts to stop it but after a few minutes has passed there came a

Blue Scars

bang and a Griffin bus had the final say, he had run over our prized possession, people ran out of shops fearing a gunshot or something that was unusual in the sound of that sudden bang but alas our ball had come to its end, amid groans of disappointment but after our initial disappointment we gathered the remains of the ball and stuffed it with paper and rags and sewed it up and had to be satisfied with that but we never collected any more wrappers after that.

So we reverted back to our usual routine of going around at evenings, taking the Argus and football echo to supplement our income, enough to get us to the pictures at weekends, the only picture houses at Brynmawr were the market hall and the Cosy Cinema and if there were any spare halfpennies or pennies would buy a Crayal (fag) and show off by smoking it while a film and try to avoid chewing gum left of the seats, my favourite films were westerns not shown today by some of the good old actors with names like Buck Jones, Tom Mix and Gene Autrey and shown in black and white in them days and not show the violence by the actors of today that leave nothing to the imagination but incite the youngsters of today to imitate the so called heroes but with disastrous results and put themselves in the role of the bullying that are prevalent today that soon gets out of control and even beyond their parents. I have witnessed myself the escalation of these events but the generation are governed by these violent scenes screened today and are copy catted. I still have memorabilia of my memorable days as a youngster and I look at the wistfully and relive them in my mind and not I concentrate of putting my thoughts on paper in the hope of them being read one day. I remember the swapping of our comics we did when boys and after their return they would come back to you with stains of tea grounds and tomato sauce but as long as the main story was visible to follow the serials from last week you were satisfied and some books handed back to you with pages missing but as long as the last pages were there to read the ending you didn't mind.

This generation will never know how happy we kids were in those days, hard times but we kids never knew also the hardships our parents had because we were too young to understand but we still have deprived areas around us now but compared to the 30's and 40's people are living in a certain amount of luxury but where you could go out and leave your front door key hanging on a piece of string inside the letterbox in case you were late coming home that have certainly vanished from the daily habits that were practised in the early days, you also missed the gathering of neighbours on one another's gates to hear the gossip of the day, that was a common feature in most every street including ours but you would have the usual bust-ups on a Saturday night, a few punch ups as well but not the viciousness of groups that you witness today, it was a one to one issue then but these incidents always happen on stop tap at a local pub or club and mostly they would fill one another a pint but you do have these little hilarious moments as I can recall one saying when we were at grub time at work, it was a man from Westside Blaina who related his true account of one of his escapades which is true according to him but can always be vouched for by the pub members. This man's name as I recall his name was Bob, I cannot remember his last name not that it matters and his account of what happened the night before he and another chap had a bit of a difference of opinion which turned out to the conclusion of a fist fight but owing to having too much to drink decided to meet on the mountain being their houses, Bob said got up in the morning and decided to confront the other man who he promised to fight so up I go and I believe the man was still a bit rough from the night before so I knocked on the door and still no answer, I threw stones as the bedroom window till he appeared and opened the window and out comes his head, his hair he hadn't combed it or he must have wore his cap to bed the state of him, 'Come on I said lets settle it once and for all' he looked at me puzzled for a minute then realized the implication of my unsocial call that early in the morning, his head disappeared for a minute and came back to the window and said to me, Bob have this half a quid (ten shillings) and bugger off, I'll see you again all right, by the time the night came and they were again at the club but things had worn off by then after being to work the both of us and the fight was never talked about so I filled him a pint, while he was drinking the cheeky bugger wanted the change out of the half a quid but anyway we tossed for it and he won and that was the end of that but Bob was a bit of a joker because one night they were pinching the coal off the full trucks in the siding his butty was always cautioned or caught and he said to Bob 'How is you never get bloody caught?' and Bob said 'Easy I see the copper coming and lay down and make a noise like a sack' Aye Bob was a bit of a spud I can tell you.

Well back to the drawing board again, months had passed by and I was introduced to a different kind of job, a different kind of turnover gang, conveyor shifting to which we were originally signed up to do but this conveyor was in contrast to the ones I knew at Beynon's Colliery, such as the shaker conveyor and belt conveyor that I was conversant with and was declared obsolete just after I left and it was a very heavy handling of the steel troughs of eight and ten feet in length which were coupled by thick heavy bolts in the shape of a curved walking stick, these troughs in turn were shaken back and fro on a wheel that ran on cradles, underneath but were controlled by a mechanical arm attached to a master trough that regulated a sudden stop that shakes the coal along the entire conveyor, this is why the shaker conveyor gets its name and I believe that the dismantling and re-erected into a new line was the heaviest conveyor to be handled because of the length and heavy steel troughs that had to be manoeuvred through wooden supports in which some of these supports were doubled in parts of the coal face because of some sudden pressure had settled on some parts of the face. I know because I was part of the team at one time, I knew that the men that carried out this task deserved a medal but it takes all kinds of structure in conveyors and to which they were most suited for in certain face but my own experience found that what suited myself my choice would be the belt conveyor by handling shorter structures. During my employment as a shifter and belts are usually in different lengths as to easier handling, imagine rolling a twenty five yard long face conveyor you could not handle such a bulk hence the use of shorter length for rolling to me being of a smaller stature found that it suited me, but when I dismantled by boots became full of the sooty coal dust that the under rollers had churned up from the returning under belt, after seven and a half hours of continuous running your eyes don't escape either by particles being blown about by the ventilation in a confined space, your shirt gets caught by the tangling wires left by the pipe fitters unable to release them because of the weight that was imposed on some supports, anyway masks were not available in them days, if they were I was not aware of them after the turnover we would retreat to the roads outside the coal face to the welcome tonic of fresher air, the cutting machine was equipped with a sort of spray but was instantly soaked up by the gumming thrown out by the churning picks of the machine and I used to wonder what conditions were by the old pioneers of a coal face and how they managed in the absences of water compared with the advancement of technology today, in them days curling boxes were used for getting coal out, they were in the shape of small wooden boxes and were dragged along narrow galleries by boys of ten and twelve years of age and women were employed at opening and shutting of large doors to allow ponies to travel to and fro to their destination with their tubs of coal. I supposed the women worked in an undignified way to procure some sort of wages and in history it was revealed that women and boys carried in bags the coal up the ladders in these shafts, what a life, what a hardship? But we have advanced not allow these hardships but not to guard against the hidden dangers that lurk in the recesses of a darkened coal mine, the only thing I knew about coal was when my mother asked to get a bucket of coal out of the coal cot, to heat the house and boil buckets of water for whatever they were needed for, but never realised the extent of the sacrifices that were made to extract the coal until I myself became actively involved but now after experiences in most aspects of mining and the friendship that develops during that time and the heroism you encounter made it bearable to end the forty two years I spent there with some of the best friends and families you wish to meet and I will never repeat the days in the pits.

At the end of the day and when you hand in your lamp for the last time you will not be able to mix with your old friends again, only at social gatherings and they will be less frequent as time goes by, but you will always have those memories behind you and to remember you by but those will fade away also and I will hang onto them as long as I can, and hopefully my biography of my life in the pits will remain as a social document, and remain as history of the part I played in Beynons and Six Bells collieries, but if I had changed my lifestyle I would never Have had the wonderful wife and children, but sadly I have lost my wife I have had for fifty eight years who was the rock of guiding my hand, never to be replaced by anyone and there my life ended to my employment, except for writing this book and to try to carry on painting of which my wife would want me to do, so I try to fulfil this part of it for her and try to revive the spark that has eluded me and to do this you have to be strong in mind and body, to fulfil part of that dream, to which I have worked hard to immortalise my friends that failed to return from the challenge of the pits, but they will not be forgotten, not by my blood anyway. My work consisted of face work that was my breaking in period, it was a madhouse when men worked in close proximity to each other which often gave cause to frustration and sometimes confrontation to be forgotten by the time the shift ended, which sometimes did not come soon enough, but that's what you can expect, you either like it or lump it.

Going to back to earlier years to an incident that will live in my memory, a mixture of fright and humour while driving a blast haulage (compressed air), it was on an occasion when a column of pipes had sprung a leak because the joints of the pipes had been hammered by the successive bumping of the wheel hubs of the drams over periods of travelling, the join in question emitted a sort of roaring sound which was a bit deafening at times, but it was a distance away and not interfering with my work, but it did interfere with the work of the hauler in charge of his pony, it made conversation impossible but did not affect me, but affected the workmen who had to pass by on their way to work, well a situation occurred while a hauler was leading his pony on the way to work. On approaching my engine the hauler tied his pony to the side of my engine and proceeded to investigate the source of the sound and while there he attempted to muffle the sound by piling bags of stone dust on the affected joint and it did the trick, I could now see him in the distance, I thought he had done a marvellous job, so he came back out and untied the pony and proceeded to go on his way and when they neared the joint by the time he had walked for the pony and walked back in the force of escaping compressed air had blown its way through the paper bags and then it started, the road way was filled with white dust from the exploding bags of stone dust, the whole road way was a complete dense of white, in which I had difficulty in seeing the engine itself, I could hear the chatter of shod hooves of the animal striking steel sleepers on the rail track and also striking the hollowness of the pipes as the terrified pony tried to escape the roar of that joint. I panicked myself I didn't know where the horse was and I had difficulty in finding a safe spot from the terrified animal, but the pony must have caught a glimpse of my lamp and headed straight for it, because it bolted for the engine house and if I had remained at the controls of the haulage I believe I would have been in some serious trouble, perhaps fatal, by the madness of that terrified pony, but luckily I managed to slide by the side of the engine and after a few kicks and snorts when the air had gotten cleaner I looked around, everywhere was a complete blanket of white, including the pony and myself but owing to the pony's harness had caught some levers of the engine, it was hanging underneath the pony but I was thinking how lucky I was because if that ever happened to be the one that controlled the compressed air that would have started the engine in motion and I was by the side and I would have been caught by the piston rod, but lucky me. So the hauler after examining the pony for any injury decided to take him back to the stables because there was no way that pony was going to pass the join and that was the only time that I was worried about the consequences that could have happened had that harness pad caught that control lever.

Back to old Bob of Westside, Blaina who was telling me that he was a hauler before becoming a timber man and told me that his pony always laughed when he was told a joke, so I said how can a pony understand a human voice to that extent, he said because horses always understand me, and the next time I come your way again I will prove it to you, do you want a bet on it? I said "aye a bob screw of bacca alright", he said "it's a bet". A few weeks went by, well I wasn't expecting Bob, but he turned up one day delivering some materials and he came up to me and said "Come on Mister bugger, I'll show what I mean", so I followed him to the horse and watched for any tricks, but I could not see one, "watch now" he said, he went to the horse and spoke something in his ear, Bob's back was towards me but I could see the horses head quite clearly and when Bob said his joke, I don't know what he said, but be dammed if the horse didn't throw his head back and laugh, his feet pawing the ground like he was really enjoying the joke. I did not say anything, I just stared, Bob said "how's that boy, didn't I tell you?" I admitted that he spoke the truth "anyway" he said "I was only kidding about that bob screw of bacca so forget it". So after he went I went and told the men, one of them said "don't be daft, the bloody horse don't know what Bob's on about and I can do the same as him, I don't want to spoil his jokes but I will tell you how it's done. While his back is turned to you, all he does is put some snuff up his nose and all the horse wants to do is bloody sneeze, that's why he showed his teeth, he wanted to sneeze all the time". "Well" I said, "he fooled me anyway I thought it was genuine" and they all started to laugh, but it makes you comfortable to have these characters about you and the warmth of comradeship that is what pit life is all about.

But the most dangerous part of face work is the work of the crashing team and I know by having some knowledge by now, having the experience of face work as much as its worth and by coming in contact with them on quite a few occasions, at their dangerous tasks and I was put to the test one week and was asked to perform as one when one of the team became absent for a few days through illness and I was told it would only be temporary. I was a bit dubious at first but the overman had been pretty good to me and that I would be doing him a favour, to which I could not very well avoid and he said there would be a couple of bob extra, so I accepted the job, but I knew my heart would not be in it. So a week went by for him to find a replacement for my own team, they had been previously asked but declined the offer, so at the beginning of the following week I met the crasher team who I knew quite well by now and they heard that I would be joining them for a few days and I asked them do they mind and they said "Come on Chop and join the party, but you will have to look after yourself because we have a bloody job looking after ourselves, and you know the ropes by now and you don't need those spanners but get hold of that Sylvester (a withdrawing tool). It was based on a lever and ratchet type with a long chain attached to it and wrapped around a steel support in the job (waste ground). It gave you a little advantage of keeping you out of danger but if the support was stubborn you had to enter the job and attack it with a sledge hammer to try and release it. I was hoping that we did not have that problem, that was the bit that I did not relish, but that was part and parcel of the job I suppose, so we all had to accept it and when I entered we all took stock of our surroundings first, myself more than them, I could hear a few loose stones dropping in the job and waited, but there was no further falls of stone and I prepared to put the Sylvester in position and took the long chain and placed it around the first support to be withdrawn. Came back and attached the chain to the Sylvester and started drawing away, the support released itself and I thought I hope the rest will be the same, so I was going in the job full of confidence and started to hammer hell out of the prop. I could see my butty "not that one, if you hit that one out how are you going to get that one out further back? You won't have any lever for your back will you?" that was one mistake I made, but the sharp eyes of my butty put me right which I should have seen myself but I thought it a bit risky going back into the job, but the one in the back had to be got out to allow the top to collapse to which I did later on with a mighty crash and a cloud of dust "that's the way see Chop", "that will give the packers of the waste plenty of muck to build their packs of stone and steady the top, and the props we have hit out from there will be used again by the colliers on the face". That was the idea of reclaiming the used props but there was one prop at the very back that I was more concerned about and that was a tricky one and it would not budge therefore one of us had to go in and attack it with the sledge hammer and I was elected to be the one, while my butty kept a tight strain on the Sylvester chain. I ventured into the waste and had a go at knocking the offending prop with the hammer, the damn thing never budged. I kept on but there was a glimmer of hope when the tight chain twisted the prop a little and now I began to get a little apprehensive and looked for an easy escape should the prop suddenly release itself, and I knew that the prop was holding a bit of weight and I was right. There came the release and I was gone like a scared rabbit just in time. I thought the whole mountain was coming down on top of us and with the down draft rush of air that followed I just stood still amongst the dust, afraid to move and as the dust cleared away I could see my butty crouched by the conveyor. He shouted "good boy Chop, that was a corker", "and I'll get that loose prop". I said "you're welcome", but if it wasn't for the use of that Sylvester I would have still been knocking hell out of that prop. My eyes were always vigilant in them times of crisis but the crashing carried on un-nerving at times, even with the experience I had at the coal face, but everything was done, it was to contribute to most aspects of the running of the coal face and the only real fright I did have was the whole collapse of the waste, that collapsed the section of the lines of props leaving a wide expanse of unsupported top, that had to be supported to be left safe for the oncoming shift and after a few unsettled moments of the settling of the unstable top, we erected the displaced props and arrested in any other weight that was expected and made the area safe to the satisfaction of the deputy who praised us for the effort we put in. My shift ended with a parting shot from one of the team "that's what it's all about Chopper, if you want to be a crasher", I replied "I haven't made any plans to be one mate", so after collecting our tools of trade, namely the indispensable Sylvester and chains and put them safe we retired to the road way and after five minutes whiff continued our walk to the bottom of the pit.

I may be reiterating parts of my story, to prioritise some of the highlights I encountered during my life in the pits and to retrain for other work would contribute nothing to society especially at my age, so this is the only way of expressing myself, my love of mining and to keep the heritage and the history of a once dominant industry.

But back to the story at hand, while waiting for orders at the pit bottom I met my brother in law who I had not met for weeks because the changing shifts between the both of us, and I was delighted to hear of exploits and his different jobs and family matters and things that were important to both of us, we were both waiting to be employed at the job we were signed up to do but that was not a priority to use at that time because the important thing to use both was that we both had a job and that's what mattered, but it would be nice to be working together but that was the way it had worked out I suppose, so this is the start of another day practically the same as any other day but most associated with haulage work which came my way very often and that needed to physical requirements, but required more thought when dealing with the safety of others and handling machinery and controlling tempers, but my thoughts often wondered to the day Dai and I would be directed to the job we were signed up for, but I thought that would come eventually but Dai was sent to help the track layers while I was sent to learn a larger electrical haulage which hauled a larger convoy of drams and a longer transport road and needed better judgement than the shorter journeys. At the beginning it was all trial and error and temperament but the ultimate aim was perfection and that everything went to plan but I relished the challenge of having trust put in me and over the years I have learned to be more tolerant. I approached my next assignment just as the driver started to operate and watched intently as he handled such a monster that had double drums that were massive, the diameter was about eight or ten feet but was half submerged in the cavity below to allow the drums to revolve. I thought to myself that I would overcome my nervousness but like all jobs you have to have a go but my weight would be the stumbling block of controlling the braking system, but I found out to my delight that the weights that were attached would counter balance even my small stature so I was alright in that sense, but I was unaccustomed to the controls because instead of levers everything was controlled by wheels, even the clutch that allowed the clutch to engage and disengage from the drums, but I watched as the controlled power was assisted by braking power, it was a lot to digest in comparison with lever control and when asked after a week or so would I like to demonstrate what I had learned, I did it with some confidence that I mustered, did in fact come up to some expectations after trying to master the butterflies that started fluttering in my stomach, but master ship would come with experience over the years so that did not present a problem, but you can learn every day and patience is a virtue so they say, and I was quite confident of fulfilling to whatever was required of me.

In-between the turnovers Dai and I before getting our regular jobs, and it would be a relief from being buggered about all the time, but before that happened we went our separate ways with both of us changing jobs and shifts so one day the deputy said to me when you finished what you have to do, I would like you to accompany me on my round of examination and he would show me mechanised face work. I said that would be great and he then said, I would not be put at a disadvantage when we would be introduced to the job we were signed up to do, so we done what we were supposed to do and when he came around he said "come on Chop", so off we go, we travelled the airways to avoid any contact with the transport of coal journeys and entered the coal face that I was used to anyway, but that was only the usual belt conveyor, till we entered the chain carrying conveyors, which were in the middle of breaking structures up. "This is the job you were supposed to do wasn't it Chopper?" I said "aye but its taking its bloody time though", "bet patient" he said, "you haven't been hard hit by it though have you?" and I agreed with that, but if and when Dai and I had been settled in one job then we both would not be buggered about the place, so we carried on and passed the packers who packed the stone packs that steadied the waste with their stone packs of debris that had fallen from the top after the crashers had released the waste props, these packs of stone would distribute the weight along the coal face to release pressure from the coal face itself. So while passing this one pack and the packer was putting the finishing touches to his pack and I thought him pretty quick but never mentioned it to anyone, but that did not deceive the deputy whose eyes were everywhere and including putting his lamp into the waste testing the area for emissions of gas accumulation should there be any, and I thought that was a good duty of a damn good deputy but he was not by anyway deceived by that pack of debris. He prodded the stone wall and shoved his stick in a hollow and the stick encountered no obstruction, so he put his foot against the wall of the pack and it immediately fell inwards. He said to the man in question "you can fill that bugger up to the top or you don't get paid, it's for every body's safety you know". I kept my eyes averted from the scene not to be laughing at the packer being caught out, but he was in the wrong, so he asked for it. The deputy said to him "you think I have come on a bloody banana boat?" so off we continued on our way up the face slowly because of packers and the conveyor structures everywhere, plus dodging the men not to get in their way, but observing the steel troughs that carried the chains, I have already described to you earlier, so I won't go into that again. We carried on up the face, the deputy exchanging a few words and a joke or two with the turnover gang, but not failing to notice any breaching of safety rules and periodically checking for gas emissions that may invariably crop up, but very rare, and that increased my respect for the good safety and the ventilation of a colliery, far in front of most other countries. I was jogged from my thoughts by the deputy saying "well Chopper have you enjoyed your little run around? I hope that you have taken notice because perhaps you won't get another chance you know?" "Well" I said, "I've noticed quite a lot including you too", he replied by asking what I meant, I said "I don't want to give you a big head but I am most impressed by the way that you were not diverted from your job of inspection so I'll give you ten out of ten"! He said, "Thanks, it's nice to hear something nice for once". We continued our tour of the face and as we were making our way out I happened to mention my brother in law Dai, and that he also had his deputy papers too and quite capable of doing the job also, be he preferred him and I to stick together. "Well" he said "I never knew that, but when I see Dai again I may have the chance to have a little chat with him". But back to reality, signing on was not quite as simple as I thought compared to the old belt conveyors that I was used to so off we went and wended our way from the coal face we had just visited and entered the main carrier belt road to where they dumped their load into waiting drams. But the deputy said "We won't travel this way because we will be travelling the main transport roads and the ropes will be flying about everywhere, so we'll travel the air way because you have enough sense for that I know". I said "I know that but you said to follow you so I have", and so we left it at that, but I was glad to have been shown the work that was going to be required of us and to have an idea of how the system works and the complications of so many parts of the chain scraper conveyors.

Dai and I had a second chance to visit the same scraper conveyor while we met when our two shifts came together and while we were transporting twelve feet lengths of pipes which required all our knowledge to negotiate the twist and turns of the supply road and the narrow confines in some of these sharp tunnels and needed all the help to deliver them to various road heads to extend the pipe columns as the road advanced owing to the travelling of the coal face, but after delivering these pipes there was not any need to return with the regular supply team, so Dai and I took a short cut up through the coal face to which we had no right to be but took a chance and watched the procedure again ourselves, while on our way up the face, joked with men in passing and said on leaving "all the best boys" and one of the boys shouted "make the best of it Chop because you will have your bloody turn when the times comes and watch where you are putting your feet on the way out, not to disturb the bolts we place at the bottom of the props, we will never find them if you scuffle them about", but as we were half way out one of the men shouted "they will be lengthening the face shortly and installing a belt face to develop another

Blue Scars

face", I replied that it would not be any good to us it would not pay the two of us because we would have to find other work to supplement our wages, but you never know what's around the corner, so we would have to wait, but we were directed later on to turn the little belt on and was made up to compensate us, but the belt had to be turned to keep up with the scraper otherwise the timber supports that covered the belt would be crushed, and it would be difficult to extricate the belt structures, so it was in our best interests to turn the belt, but we complained so often that there was a note on our lamps that we had to see the manager, so off we went and as we went into the office the manager's clerk said "what have you done again?". I said we haven't done anything but we entered the office when told to enter and were asked why we were refusing on occasions to do the job. I replied that we were doing the job alright but under protest because of time we had to remove the structures and would not have time to make the other jobs pay our piece work. After a few minutes he said fairplay you have not been any trouble and when asked by the deputy you have always done what was asked of you, alright he said, I see your point, but leave it for now till we sort things out, after a few days went by we were sent for again and was explained that the whole team would be worked together and that we would have to work through and through with us helping the scraper gang and would be put afternoons the following week and that was a start of thirteen and a half years of afternoons regular. I didn't mind it in the winter because when you are down there it's all the same, except for pit bottom and we were not working anywhere near there, thank God, so I can imagine what them chaps were like I'm sure, but in the summers we used to have, they were glorious, we don't have those summers now, but when we

Chopper Davies

had those summers it used to break your heart, we would be having our fag on top of the pit and see the men coming up off the day shift, they would make it worse by throwing their helmets in the air and looking at us and pointing to the sun, did get to me the first few weeks but after a while it wore off but after a fag or two and stepped on that cage and the sunlight finally disappeared, it was back to normal and soon forgotten about and resigned to seven and a half hours of slogging and the only thing I missed most of all was my couple of pints in the Golden Lion on the Garn but you cannot have everything I suppose, but it was our own fault for causing a fuss about our job, I have no-one to blame for the predicament we were in but we could not take the risk because I now had two children to support who relied on me now and they came first and our increased wages were a good incentive too, so you had to take the good with the bad, anyway I'll stop moaning and get on with my story.

So on entering pit bottom we presented ourselves to the officials and explained the new arrangements and were told by them that they knew all about them, one of the deputies did say though 'Are you bloody satisfied now Chop'? I said that is what we were signed up for anyway, so off we went on our way to meet the rest of the team who were having the usual whiff before the mad house started to get underway, the packers of the face had already started so Dai and I had a few moments more to sort out what we required, the old favourites the Sylvester and chain, because our job first was to turn the small belt conveyor into a new line before we joined the rest of the team in dismantling the scraper conveyor and after being told the routine which was not too hard to follow, but far from the handling of the lighter structures of the belt, but after a while the team was

Blue Scars

cemented together and adapted to the job, you had to get used to different attitudes of the men who like myself had our own individualistic views and temperaments, but myself I had no special qualities other than try to be as pleasant as I could owing to the newness of working as six of a team where as there were only Dai and I so that had to get sorted out and expected to get a few hiccups in the process, but everything went to plan.

But back to my time as an eight or ten year old in 1928/29 era, I reminisce back to those days but most of the old characters have moved away by now, or passed away I suppose to which undoubtedly I would be one of them but not yet at the time of writing 2005. The brick and stone buildings that were about in them days did not need all this loft insulation that are about today with the precast concrete panels that are erected, the warmth was always there in the old buildings when coal was available, but you have to go with the times I suppose, I don't know if it's for the better, I was always used to the old black leaded monsters of grates, to be replaced by imitation marble effect grates today which enclosed the good old roaring fire with a glass fronted unit, that don't give the homely atmosphere that was given then. I still retain my coal fire, but like I said doesn't have the same effect in the cold winters as we are experiencing at this present times, most homes have the conversion of gas and electrical appliances that are met now by spiralling of the cost of heating, but many pubs still keep to the old traditions, at least the pubs I visit now I have retired due to ill health, and not many miners leave the coal industry without being left with some legacy of ill health and blue scars that will be with us the rest of our lives I fear, including myself. But what still remain are the traditional games in the pubs and clubs, such as darts, games of cards such as cribbage and nine card Don and various games, but at the sound of the bell at last orders everyone were loath to leave the warmth of that fire, but I will never see the price of a pint at sixpence again or the price of a packet of Woodbines at two pence and four pence, the prices today are near enough £5 for twenty Woodbines, I wonder what the price would be in the future, but people will still buy them if affordable because people will never give up the craving for a fag, I for one of the, it is the only pleasure I have. But at home I had my chores to do, even as a youngster, by helping, well expected to help, when the tin of Zebra polish was brought out to bring our grate to its shining glory, to stand out like a gigantic sentinel against our dresser of plates and cups and the scattering of family photographs and letters wedged beside the ornaments all this bring back nostalgic memories that can never be replaced, perhaps a good many would say (a bloody good job they are gone), but I remember the clock on the mantelpiece to which my knowledge had never kept the right time and stood like guards alongside were the two china dogs, my father's watch hanging on a nail by the side of the fireplace covered by a brass case to prevent the dust from the pit, it was good to remember the array of brassware that surrounded our fireplace with the toasting fork in its usual position beside the watch, but I think today would be a drudgery of polishing brass pieces, they prefer china dolls and glass ornaments, but the taste of roasting potatoes mingling with the spewing out of that blue gas from the coals makes the food today tasteless, we put up with the burnt tongue and blackened lips from the charred skin of the potatoes, the only thing that came out of these treats was the unfriendly smack across the ear when wiping your hands, thought to be unnoticed on the towels, which was

Blue Scars

not very hygienic I know, but when frying something on the open fire the added flavour given by the smoke give you a further appetite, especially and beans. Them days sides of bacon were hung outside butchers shops alongside of other carcasses, fish were openly displayed by an open window, so was other commodities like cheese and green produce were left out in God's given air, but today you are faced with deep frozen foods filled with additives and foods that probably have gone by the sell by date, people don't seem to notice, or probably in a rush to do their shopping, the microwave will do the rest, with their bombardment of atoms, but you haven't much of an option today, perhaps things have moved on for the better but I fail to see it, all you can see today is 'junk mail' coming through your door or bills, bills, bills—for Council tax, gas, bills for electricity—that is the advantage of coal, that cuts out these out of proportion demands, coal fires had numerous uses like boiling the kettle, heating irons on the red hot coals to iron your shirts and sundries, the iron was on the table top and the irons when sufficiently heated were spot on to make sure that is was sufficiently heated and wiped with a clean cloth, but what mystified me was the fact that all the items that were ironed never left a smudge or mark and again the shirts were as white as driven snow by the fact in them days a cube of Reckitts blue was put into the wash, a blue colouring put into a cube exactly the size of the cube that billiards or snooker used, the only drawbacks were when using starch for the collars your neck was bloody well raw because the majority of shirts them were collarless and you had the difficulty of entering the studs into the slots in the collars that secured the collar to the shirt. I used to keep my studs in an egg cup on the dresser but on occasions you happen to mislay one you would have to go to the shop and get a card with two on it because they would not sell separately, but even when the collar was on your next, when your tie was applied you had hells work to turn your collar over the tie and required the help of one of the household, but my mate one day lost his one stud and the only way he was to come out was to use the little nut and bold that secured the rear light of his bicycle and by the end of the night left rust marks on his neck and collar (and this is true) unnoticed by him but was cause for amusement to the rest of the gang, but when I was working down the pit I used to see the oldsters with their collarless flannel shirts and braces holding their trousers up and I thought, bloody hell I would be scratching all day with the flannel rubbing against my skin but I assumed that they always wear those flannel shirts.

But everything turned out for the best after all as a working party but you cannot foresee ahead but our arrangements went on for quite a while, in fact a few years, till eventually the team got whittled down to four of us because the other two had reached retiring age and ill health, but by that time the short belt conveyor had finished its useful purpose of lengthening the face but we were sorry to lose our friends, who were good friends, and were as good friends as you wished to meet and would be sadly missed because in the time when were working together we knew more about each other than we knew ourselves, we knew each other's moves to a fine art but they got out at the right time before the problems started to develop, because water problems appeared eventually giving cause to management and ourselves, but more for us, because we had the trouble to deal with it and to work in it plus the undulations that appeared in the floor bed, which were called differential subsidence, where one part of the floor subsides but the rest

Blue Scars

of the floor stayed at its normal level but in these undulations the problem of collections of water which had to be dealt with by the introduction of sturry pumps to drain the water, but the sediments left in some of these depressions, it was that made our work uncomfortable having boots filled with a mixture of water and sediment and the retrieving of some of the structures while kneeling, so you can imagine what we looked like at the end of the shift and who would want the blasting of released air every five minutes from the damn pumps, so I could imagine the work of the colliers at their tasks, so these are some of the trials that confront colliers and the workforce on faces that have water problems and to make matters worse when machinery had to be replaced and dragged along the face for these had to be handled with caution owing to the slippery mess of oil and water from the transport o fit in the supply journeys and which had been standing under sodden timbers that dripped from the ring arches in the supply road, the only protection we had was to empty stone dust bags and make a hole for your head and arms and cover yourself in that way and try to imagine our shirts under the caking of the contents left in the stone dust bags but that was the lesser of two evils, but as the face advanced our trips to the boiler house at the surface became more frequent to dry out wet clothes and bang the against walls to get the cakiness out of your shirts, but it did not deter us from taking the item required and deliver it to its destination.

There have been quite a few changes in our valley since the demise of coal and steel, leaving resentment against the Government of the day, namely Thatcherism, it caused emotion and effect on local communities that depended on these two great industries to provide alternative employment for their members, but we have to look forward in the absence of them, but the scars will always remain, even if memories fail they find solace in the fact that they did not surrender without a fight, history will judge those moments and history will not forget the miners who fought vociferously to defend their rights, but unfortunately, the Government of the day buried our future to the disillusionment of many who were employed in both coal and steel but out of an empty stomach there always springs a leader and consequently that happened by the stubbornness of the miners at Tower Colliery who had the guts to take a change of buying their own colliery with their redundancy money, that's guts, but it still stands as a monument to valleys fight and still producing coal and a success to that body of men at Tower Colliery, I take my hat off to those men, but many men find solace in the drinking habit in which they had no time to hang around pubs and clubs, except when they finished a day's work, but now they have time on their hands smoking has increased, as youngsters develop a sense of boredom creating new challenges to the unwary, like the challenges of drug taking, but where smoking is concerned I cannot throw stones at anyone regarding that habit because I also started in my early stages of my life, but intimidation by the older element and under threat if I refused, but my first introduction was in the form of a Woodbine to which when I inhaled the smoke subsequently became sick to the amusement of the older boys. I can still remember the names of the older cigarettes, namely packets of, Four Aces, Black Cat, Kensitas, Star, Loadstone and could mention many more, in fact I am still smoking today in my eighties, but it's nothing to brag about I'm sure.

Blue Scars

But at Six Bells all things come to an end and some things turned out for the better in some ways, you have these runs o bad luck and good luck, but after our team had been whittled down to four of us, after the retirement of the other two and as the time passed by conveyor shifting began to take its toll on me physically and I wondered if I would be capable of handling heavy structures, and the pressures demanded of you to body punishment and Dai and I were talking about this and I suggested to him that we were getting on in age and that if I had his qualifications I would be thinking of making a move to a more secure job, such as in his case, taking up a deputies position before it was too late, there was not much chance of obtaining that prospect in our line of work, or in any other line of work in pits, only hard work to which we were both afraid of, but to look at it realistically because age don't come by itself, and after a few weeks went by he started to consider our talk but was reluctant to breaking our long relationship to which we had worked together, but I assured him that I wasn't going to keep up the job I was doing at the present. He could see my reasoning and that he may be leaving too late the chance of bettering himself so he made a decision in the best interest and went for it and asked for an interview with the manager. I don't know what transpired at that interview but later on in the day he told me that the meeting have proved favourably to Dai's request and waited for the results. Dai then presented his credentials from his former colliery Roseheyworth colliery and after a while was accepted as an official, the manager wanted to know why he hadn't presented himself before, Dai said we started together and did not want to break up our relationship on principle an of being a family member. So after a few months went by and the scraper conveyors use was being taken over by power loading machinery that did not need turnover gangs, so that at the end of the scraper era power loading was the main contributor of coalface working, in which the coalface was controlled by hydraulic conversion with better roof covering that outdated prop support, which was in the shape of steel canopies that resembled giant ironing boards that were use in the homes and the structure of the conveyor was controlled by hydraulic pushers that moved the entire length of the conveyor, tight against the coalface, the use of the conventional cutting machine that used horizontal jibs with picks were more or less obsolete compared to the power cutting machine that sheared the face wall and a plough that was attached would shove the coal automatically on to the conveyor and whisked away by the flights that ran inside the channels, and instead of picks and shovels they used small poker like pins to go into the holes of the hydraulics to control the movements of the machinery.

So Dai and I went our different ways, after the acceptance of his new job and I was left in the usual way of being buggered about again. I lost sight of him because his job required him to work three shifts, days, afternoons and nights and would only meet occasionally when I was turning small belts over and showing new butties in the turnover game but most of them were not interested in the job half the time and I could have done the job myself in half the bloody time, by the time I showed them what to do, but the most annoying part was that I wasn't paid the yardage they were getting, after working hard in showing them what to do under my supervision, that is when I missed my mates but I realised that I was placed in a similar position myself at one time but I did at least take an interest in my job, but as the little developing face progressed the boys did show a little more interest but I was called in to the face should they encounter any difficulty in alignment at the head and on one occasion I had a surprise visitor while trying to align the head of the conveyor, it was none other than Dai on his round of inspection, he said 'What's the taff'? I said 'The usual thing but don't worry the job will be done'. 'I know that' he said 'Because I'll give you a hand' and so he did and he reported that the belt was in running order. I met him once after that while repairing a broken rail track and we sat down together and reminisced about our past experiences and how things have changed, chatted about the family and off he went, it was not long after that he was transferred to our next door pit which was the up cast pit that received the ventilation from our airways, I never saw him only at family gatherings or we would meet at a pub or club. Si I got on with my work as a spare hand conveyor shifter, of which I was never out of a job, but it came one time on a Christmas time and I was sent to a small conveyor about sixty yards long and as it was near to Christmas I had all the help in the world because of the colliers rushing to get their coal off which they did a bit early and was ready to help in getting the belt structure out, but6 they did not care where they threw the belts and attachments and I thought it was good of them but look at the bloody sorting out that met my eyes. I believe it would have been better left alone to me, what a shambles when I entered after the men had rushed out and I didn't know where to start and I was on my own now, the face was covered with rolled up belts, the colliers had not cleaned the belts before rolling them up and they looked the size of a steamroller in such a confined space and the supports were doubled because of the inactivity over the Christmas period, it was complete carnage, in the colliers scrabble to get out after their coal was off but now they have left it was as silent as a tomb and when you are on your own everything seemed magnified, even the falling of small bits of debris falling on the empty structures made you jump around, so I went back into the roadway to settle my thoughts to make a start, then I saw a light at the end of the roadway and as it came nearer someone shouted 'Everything all right Chopper'? I shouted back 'I'm OK' and as the man approached I could see he was a chap from another district who was sent to stay as near as possible to the road head and made sure I was in shouting distance because it was unlawful not to be in shouting distance in the pits, so I was comforted with that, the man said 'I haven't come to help you Chop, only for you to have contact with somebody should you be in difficulties, that's all I am here for if that's any help to you'. I assured him that I was quite happy with that and I asked him why he hadn't gone home. 'Well' he said 'I am living on my bloody own so I might as well be down here with you, but I am not coming in there that's for sure'. I was grateful for that anyway so I went back under the lowness of the face and got stuck in. I don't know that I would have been able to complete the job because of the disarrangement of belts and structures, the other worry I had, would I have enough life left in my battery, because I would have to work over my own shift. The time had gone so fast that I had not noticed the time till the voice of the chap in the roadway, 'I'm off now Chopper to catch the cage, I have had a couple of hours with you now, so I'm going and look after yourself old butty, the deputy will be around shortly anyway'. So after he went I went and had five minutes whiff and straightened my back, after the whiff I went back in and worked on steadily laying the structures of the belt. Because I was on my own I work more steady because I had to keep my eyes everywhere, I was so wrapped up in the work that I did not hear the

Blue Scars

deputy crawling up the face until he was quite near me and when he shouted I very nearly had a fit 'Are you all right Chop'? What I called him was nobody's business but I was certainly glad to see him I can tell you if it was only to have an extra light because now the weight had really started to settle on the supports and the face in general and I started to put a move on because of the continuous thuds of fallen stone in the waste plus coal slips detaching themselves from the coal face, but my job was very near completed now, except for the rolling of the belts and setting the tension box at the end and attaching the Sylvester's for when the time came to put the belts together after pegging the motor, and then the tensioning of the belt to put it in running order and after setting the belt and tensioning end, the job would be finished and that would be a relief I can tell you and then I would be ready for home so after having the help of the deputy to join the belts and ran the conveyor for a while to make sure that the alignment was right at both ends with the belt running true, that was that, and made my way out to the road to straighten my back, had two minutes whiff and I was putting my coat on, so the deputy said to me, have five minutes Chop for me to see the rest of my round because I have been held up by you, I thought that's the least I can do for him after he had given me a hand so I said 'Righto' I'll have a whiff for you', half an hour went by and he returned and had two minutes and finally broke the news to me, 'I have been on the phone to pit bottom and reported the job was completed and in running order and they were satisfied, but they put me in a bit of a spot Chopper' he said. I told them everything was all right and they haven't any problems so why worry, the reply came back, we know all that but the crashing team have not turned up for work and we cannot afford to put anyone in their place and we can't afford to leave the props behind in the waste because we won't be able to retrieve them over the weekend owing to the weight being put on them, ask Chopper will he try and crash the waste and you stay there to help him as much as you can and I will send an extra lamp for Chopper. I said 'Don't you think I have done bloody enough for one day and I am not taking the Sylvester's off the tension box now that I have tensioned it up'. 'Well' he said 'if you agree to do it we will have to borrow the ones of the power loading face and I will go and get the extra lamp and if I've got to I'll work with you to do as much as we can, is that alright'? He could not be fairer than that so I couldn't very well refuse, so I agreed to do as much as I can, so it was settled that I borrow a Sylvester from the power loading conveyor but I was a bit squeamish about crashing because it was some considerable time since I did that and I wasn't sure how the deputy could help in that respect but I had promised and I would not be on my own anyway. So while he went to get the spare lamp I made my way to the panzer face to borrow the Sylvester and when I reached it there must have been a maintenance gang at work there, about a hundred yards up the face at the top end and the lights were shining in my direction and it enhanced the surroundings of the wet face, it was like looking at a scene from a film of outer space, I cannot describe or explain the symmetry and arrangement of the regularity of the wt flights that carry the coal and the hydraulic props stood in regularity also and stood as if they were sentinels in wet armour, guarding, it was awe inspiring to view in silence as the heavy wet chains disappeared in the distance, so I dragged the Sylvester over obstacles that lay about, littered with electric cables that supplied the power to the panzer conveyor, the cables looked like giant snakes with their shining length of wetness. I was awakened out of thoughts

Blue Scars

by the shout of the deputy 'Come on Chop' so we carried the Sylvester and chain back between us to the belt which looked like a dinky toy to the monster we had just left, so we got on trying to extract the props from the waste, at least I did because the deputy was well back, but assisted pulling the extracted supports out and placing them in a orderly fashion by the face conveyor, he gave me plenty of room to get away should I sense some danger fair play. Eventually I had gotten half way down the face and I had to have another whiff for five minutes but I could not rest until I had completed the job in hand, to the delight of the deputy and we retired to the safety of the roadway and could hear bumps and thuds as the parts of the now unsupported waste started to collapse, I was glad that we were out in the roadway, I asked the deputy what about the Sylvester I had borrowed, he said 'Don't worry about it I will leave orders for the men to pick it up on their way into work, you have done enough these last few hours and thanks for helping out, things will be put right for you don't worry, I know it's not a lot of thanks and not a very good start for Christmas' Anyway I wanted my kids to have the same as everybody else's did. 'Come on' the deputy said 'Let's bugger off' and that was that, we made our way to pit bottom and up the pit, thank God for that. So I had my few pints and a good Christmas with my good wife and kids, I was satisfied to go back to the grindstone after all the excitement of Christmas and I loved every minute of my work down under, coal was in my blood now like every miner that knew no other job apart from the pit and I haven't seen many leaving so it cannot be as bad as all that.

In the 1930's we used to have our own assortment of characters on the garn, including myself later on I suppose, but every village and town had their own in giving some form of entertainment, but some were on the border of idiocy and some residents did not know what humour was, but some characters were the subject of ridicule and were made fun of by us kids on the block but were more intelligent than we were. They were not so daft as they made out to be for instance they would do most things as long as there was a pint in it but apart from them if everybody had the same ideas and attitudes as everybody else I think it would be pretty monotonous but I would not attempt some of the antics as some of the characters who specialise in their own way of entertainment. There was a particular chap who had his usual way of establishing himself by way of doing a stunt of performing by falling backwards off a chair into a tin bath of water. He would start when someone would fetch a bath of water and then he would strip to his underclothes to the encouragement of the men in the pub. He would stand on a chair and fall backwards into the bath of water amid the applause of the pub regulars who would then fill him with a couple of pints so that was one way of getting a few drinks. His name was Herbie Rooke so they tell me, but most of the time he would prefer money so that he would be able to continue his round of the pubs. He was seen to be doing this kind of act in the Fireman's Arms pub in Garn Cross area, opposite the general post office, but the pub has been closed a long time to make way for flats. Another character on the Garn was a chap by the name of Ozzie Lovat who was a little backward and he was the resident character who was never without a cigarette in his mouth which was never dislodged even when coughing his lungs up, that fag was a fixture in his mouth, always around six or seven in the morning he was coughing and cursing oblivious to anybody that was in the vicinity but nobody would ever question his cursing because he would give them their answer in no uncertain terms, so he was left alone. One of his chores was to wheel coke into the local chapel sheds, Hermon Chapel (but now demolished to make way for a doctor's surgery), and fair play he did a genuine job at that by barrowing the coke into the chapel, but one day the vicar who lived adjacent to the chapel whose name was Cwm Toch, well the vicar failed to pay him and Ozzie took affront to that, when the vicar approached him saying that the lord would pay him, I cannot repeat what Ozzie said to him, but I will leave that to your imagination of no doubt you will, but Ozzie did say to him that the lord can put the next barrow load in his bloody self and people passing by took no notice of him, afraid of having a mouthful back. Well he worked for the council and if he didn't like you while being employed as a dustman, he would throw your ash buckets onto the lorry and you would have to run after the ash lorry to get your buckets back. We always used any type of container to hold our household waste such as an old tin bath, biscuit tins, but it was no good to argue with the likes of Ozzie Lovat, you had to grin and bear it so you had to be social able to him and that was easier said than done.

I know if I left the pits I would sadly miss my mates who I had worked with for years and known all my life and I believe that when a miner hands his lamp into the lamp room for the last time he loses his identity because you cannot meet these mates as we all get dispersed over the years and get spread all over the valleys and perhaps you would not see each other or make contact again so you rely on your memories. I never look back on bad times, I always look at the good times we all had together. It's a great pity that we can only meet at some social gathering or the trips when the miners took the yearly holidays when the mines had their annual holidays, but even that has vanished and put paid to. Like I have stated Tower colliery had the guts to take on the then conservative government and won, more praise to them but this is my biography as a tribute to all miners wherever they are, this story is straight from my heart until I was made to retire through ill health in 1977, but I have been very fortunate in having only a few broken bones, buried a few times, but have in some respects been very lucky after forty two years, but no miner leaves the industry and I have witnessed a good many broken hearts that will live with me the rest of my life.

I started back after Christmas to a different coal face that I had left which would have been worse if the deputy and myself had not withdrawn those waste supports I can tell you and mother nature did have her way after we left because the face supports had strewed in all directions caused by overhanging coal collapsing from the coal face forced by the weight pressure, water pipes, blast pipes, were left dangling from broken wires, that once held them up, while others were buried under an assortment of debris, so the cutter track had to be cleared and put safe in order that the cutter could continue with its job, but looking at the face at that moment it appeared a no go area and seemed a little frightening and I did not envy the colliers putting things in workable order. The drive motor at the head of the conveyor had lost its pegs holding the head in position that was my problem now, which obviously I could not possibly do on my own so I had to acquire the help of the belt attendant and he was not at all pleased at my request for his help. By straightening the head and re-pegging it I came under criticism for not putting stouter pegs on the motor and I told him had I done so I would have probably busted the gearbox which could not have withstood the weight of the top, thus the breaking of pegs released that weight having put that right proceeded to the tension end, and finally had the conveyor running as normal and bugger them I thought get on with it, but I was amazed at the enormity of damage that was caused by a week's absence and wondered what would happen if we had not done what that was expected of us (the deputy and myself I mean), before Christmas, but everything went according to plan, grub time came and I listened to the exploits of what happened over the festivities and how much they had spent on the kids and the beer they had consumed and I heard one chap saying that he was bloody glad to come back to work because he was busted and I heard a couple saying that they were fed up by the end of the bloody week anyway and glad to get back to normality, and I heard one of them saying and pointing in my direction good thing Chopper got them supports out of the waste or we would be looking damn well in our stents (place of work), "good work Chop".

Looking back over the years of the 40's and 50's we saw a thriving community, you could go anywhere on the garn at one time and stock up your larder with every kind of vegetables, meat, pigs head brawn, dripping pork or beef and they were fresh, no supermarket rubbish that may have been kept frozen for weeks and past their sell by dates, but could kill you by freshened labels and the stomach churning smells that emanate from some of the ingredients and additives that are added to make some of the concoctions, the scientists today will tell you what to eat today, but tomorrow tell you a different story, conflicting themselves when a few weeks have passed by, but in my day you did not have to go out of town, except for things like clothing or something that depended on what money was about then and what the fashion was about, apart from flat caps and scarves, but I was happy with what was handed down from my brothers, like taking a bit off the bottoms of the trousers and other little things that would make do, we had our own ironmonger owned by a man called Mr. Green who had most of the things you needed, like buckets or brooms sprigs (small nails for your shoes), studs of every description and even had a cycle in the shop window at one time I can remember. But he wasn't much of a social character at least I didn't think so, for I was too young to understand I suppose. We also had a greengrocers but the main one was on Garn Cross owned by a Mr. Fawke. We had our fish shops that provided our staple diet, one which shop was around the corner from our house in Vincent Avenue, owned by a Mr. Jones and family, we had barbers, chemists, cobblers (shoe repair). I can always remember one cobbler in particular because of his habit of having a mouthful of sprigs to nail the leather soles to the boots, his name was Mr. Hale. I often wondered did he ever swallow any of them, I expect he did, we had the local doctor by the name of Doctor Verwey, incidentally the street I now live at this present time was named after him, and they were more accessible in those days and did not have a queue like going to the market hall. I don't know if there were less ailments in those days or not, but I would not know, except that I notice they have to employ more doctors now more than ever, it makes you wonder where all these diseases are coming from. Like I said, half of them were not about then I suppose because I was lucky or perhaps I did not notice, but it makes you wonder of what the food you are eating, but if you wanted to see the doctor today you would probably need to give him a week's

Blue Scars

notice. No fault of theirs because of the increased pressures he is under today, fair play they are not bad.

Most important of all are the post office at Nantyglo has not moved its office since before I was born I suppose and still stationed at Nantyglo and still there, near to Garn Cross in King Street, but I have seen quite a number of postmasters and postmistresses in my lifetime and anyway it is one of the best places to meet people and have a crank (gossip). There is one place where you can really hear of the day's events and that is at the local barbers shop, if you wanted to hear anything locally I know it was only a four bed roomed house, two up and two down. Downstairs consisted of a kitchen and little side room but a lot of business was transacted there but mostly horse racing and the betting that goes with it (undercover of course), especially pigeon racing, but naturally took second place to horses and dogs and that tiny room was always full of smoke from fags and pipes, and it would sometimes burn your eyes when you entered from outside and by the time the barber went and boiled the kettle to get hot water for shaving he took time to have a fag himself, plus cutting the Argus into squares to wipe the lather of his razor (cut throat razors they were called then because of the open blade), you could hardly keep your eyes open by the time he prepared himself to start his shaving and after a few hacking coughs he would proceed, the barber was a quiet and nice man whose name was Emlyn Rogers and I can remember the two marble basins, the full length of the room and one was cracked anyway. I don't suppose they would hold any water and only there to show a bit of character to the place and to receive the bits of paper he would throw in there after their use in wiping the blade, and Emlyn was sadly missed when he passed away (Good old Emlyn Rogers).

And another of our local landmarks have also disappeared it is that of Harmon Chapel built in 1820, that dominated the square at Garn Cross, to be replaced by a doctors surgery and above the chemist next was called Victoria Rooms where bible classes were held for us kids on a Sunday morning and gave lessons on the scriptures, at the end of the session we were handed a paper photograph of Jesus Christ or his disciples to prove that we had attended our lessons although its disappearance has left us with a void that cannot be replaced. Memories will always surface by photographs that are still retained by many of the chapel goers and the history of local buildings and the other remaining landmark is the big white house that dominates the centre of Garn Cross that have braved many battering from the weather that has gone on over the years and whose windows have reflected a many happy faces of different kinds of parades, especially school walking days with their brightly coloured dresses, bonnets and the usual bands that accompanied these old time processions, the crowded pavements filled with sightseeing local and out of town parades, but you can always be sure of the Icon of the special days, the old favourite the Ice Cream Man, if those windows could speak it would also reflect the hardships as well, the old co-op building is there but after a while it fell into the hands of Asian people who lives on those premises. I can remember the goods being deliver by the horse drays that delivered the goods to customers, the ones that were lucky to have goods delivered and as a kid I used to watch the money changing system, the spring propelled little carriages that shot across the room, the changing desk on wires suspended

Blue Scars

above and returned the change to the customer, but the atmosphere by some of the employees were different from the checkout girls of today because there were a sense of haughtiness by the staff in them days and a sort of strictness about them, compared to the cheerfulness of the checkout girls who mostly had some comment to make and a sort of friendliness about them, that is what I found anyway. But a few girls had to go to work in service to get that extra money to help in the house by sending money home and could not come home on weekends either, I know, because my sister was one of them. We had a greengrocers on the Garn Cross owned by Mr Fawkes and was a stopping place for the buses, Ralphs and Griffin buses were the dominant bus services in them days, Ralph buses were based at Abertillery and the Griffin buses based at Brynmawr and in the doorway of the shop was installed a time keeping machine to register the arrival and departure of buses. We kids used to wait for the key to come out of the machine to have our hands stamped by the ink on the key, a few of Ralphs buses had ladders attached to the back of the bus for access to luggage that were placed on the roof rack and it also provided us kids with a ride but the more kids the further up the ladder you had to get, the least chance you had of getting off by the time the bus reached the next bus stop and woe betide you should the conductor have caught sight of you through the back window. We were petrified as the bus gathered speed and no one got off at the next stop (passengers I mean) then the bus would carry us to Brynmawr, it was stomach churned of being caught by that bus conductor who would try and kick you off the ladder, but we did not do that very often I can tell you, but it was a thrill sometimes.

Chopper Davies

But it is not the end of my coal face experience, far from it. I was asked by the over man one day, would I help the cutter man because his butty hadn't turned up because he said you have enough knowledge of the work by now and would be more suitable to help him and as my turnovers were few and far between I agreed, it was not many jobs I have not done while working at Six Bells but I was never a cutters assistant before, but was quite happy to have a go, but still available for the turnovers when or where they turned up and I knew I would have to call a halt one day, I suppose, I could not go on forever but it happened in my case later on in life but so far I have escaped any major accidents during my years at the pit. Anyway some of the coal faces have been abandoned, some with water problems and others have come to end of their boundaries but there is always something down the pit, but I have given you account of turnover gangs in general, I Have seen a few fatalities in the pit that would have deterred many but mining was firmly in my blood now and always will be and would take some shaking off, even if you had to rescue men from awkward places but that can be expected when working in those confined spaces, it is an occupational hazard.

So let's hope that my contribution be a tribute to my fallen comrades that I have worked with for so many years and hope that this book, if ever published, would fall into the hands of someone who will appreciate how I try to immortalise these pioneers of the mining industry and it is they who built the foundations that laid the building of our mining communities that stand today and I was proud to part of and still are, but at the time of my writing in 2005 the demise of the coal industry was already decimated, despite the valiant efforts of everyone it proved to no avail, except

Blue Scars

for Tower Colliery in which I visited on a few occasions and in fact presented them with two of my paintings and hung in their committee room to the delight of Tyrone O'Sullivan and Co directors but I can assure you that the pleasure was all mine. But lest we forget my warmest tribute goes to the womenfolk of the valleys, that without their support they could not have survived for so long, without their charity shows and collecting that kept the spirit of the men folk and moral of the strikers, I speak of our women in the most glowing terms and their fight will go down in history but we still retain our humour, when one chap answered an advert for a holiday that said it was five minutes from the sea, but they didn't tell me he said that it was by phone, but myself I don't go out a lot anymore, I continue with my obsession for painting and try to put my thoughts into pictures from to which I have attained a certain degree of success because pictures can speak a thousand words. I myself have collated approximately 400 works of art so I wonder how many words can be spoken by them, but the pictures also emanate the strongest of one's feelings, I may have stated previously that I am not a Lowry or a Picasso but I put the same passion into my paintings as they do, because all artists put 100% into their work, but one painting of mine has a particular residence in my home and defies an explanation, it involves the appearance of images of miners, trapped miners, in which I have not any knowledge of painting, so something or someone added an extra 20% into this phenomenon which defies TV and national newspapers but eh picture itself must have been meant for me. I have had visits from HTV by Rebecca John, reporter and broadcaster, at my home for a period of six hours with her cameraman, it has been introduced by Sarah Edwards on TV Wales Today and narrated by Geraent Vincent news reported, but even the amount of publicity it has received it has revealed something that still mystifies everyone, even when I was interviewed by Roy Noble on his daytime talk show, the proof is still in my possession today and I am proud of it, it shows what it does show is of my commitment to our heritage, it seems that a gift had been handed to me and I intend to take full advantage, it is said of strange happenings that have happened in some of the old pits in the South Wales valleys, could this be one of them? I myself believe it and the paranormal and phenomenon activities that were present in some of the pits in the South Wales coalfield that is why I keep my painting separate from my other works.

All this could not have been done without the support of a good wife who supported me all my life and whenever I was visited by these celebrities she was always at hand, in the case of Rebecca John she made a great impact and by many visitors but plenty of welcome was the order of the day and I wonder will with all the efforts we both put in, will it ever bear fruition one day, but we have left a legacy behind us, for both my children David and Elaine have families of their own now but we have eleven great-grandchildren at the time of writing 2006 so don't know if they will take up my ambitions but there again we have a different generation today, but since my wife Tilly passed away in 2002 the spark has dimmed a little and I am doing my best to rekindle that spark into a flame that it was, but still try and be a good father as my wife would wish.

Back to normal again, as I continued my work as a turnover man but I had to have a butty on this occasion because of the face had been extended to greater length of hundred or more yards and would afford us a decent wage and it suited

Blue Scars

all parties concerned so I wasn't worried but the shift pattern changed to ten in the morning to half past five in the evening, we called that shift (bacon and egg shift) and that went on for a considerable time, by butty and I went down at ten and cleaned the waste side of the conveyor in preparation for turning because it was a mess of spillage, thrown about by colliers to get rid of their yardage that was allocated to them. I believe they threw the coal over the belt not on it to attain that goal, well I could not blame them because I remembered an occasion when I worked in Beynon's as a kid I did the same, but the shoe was not on the other foot and the turnover would have to be completed by the end of the shift, but before we started our shift, men on their various jobs all met at a point near the coalface, there was Wally Davies the borer from Brynmawr whose father kept the Rising Sun and Brynmawr, whose nickname was Tiger, there were a selection of timber men and we all chatted to what we needed on our jobs till it was time to enter the face. I collected my usual tools, well one tool anyway, it consisted of a shovel with a split down the middle and no tip to it but it served its purpose, you did not need a special tool for what I had in mind, I had already had my main tool in its holster on my belt and that was my main too, it was a pliers to pull the pins out of the belt, separate them in their lengths, I had made an alteration to the pliers by placing the levers between the joints in the rails and bent them to form a curved grip, so should the belts be wet and slippery then I would have that extra grip by pulling out of the pin and I did not envy the borers tools which consisted of coils of hose that would be attached to a heavy shoulder boring machine, but they also had to carry on their shoulder a hydraulic leg to raise the machine to a required height and I can assure you I had the lesser of two evils, but I can see

Wally and butty in water and oil soaked clothing, with the leaking of some joints in the hose lines. So I went on into the face amid the roar of the belt engine (unit) and flying shovels and blast picks and commenced cleaning away the small coal debris which incidentally was used by packers to erect their packs which should have been filled with stone debris to make the pack more solid and resistant to roof weight but I wasn't concerned with that as long as the belt was uncovered, but after the departure of the colliers I noticed that there were still bunches of coal still left on the face and that had to be got rid of to make way for the cutter to do its work, but not only that, I would not be able to lay the belt structure down the face. I thought that's buggered it so all I could do was to lay the belt as far as possible until the coal bunches had been removed and that meant that the night shift would have to get rid of those bunches, that meant again I would be doing something that I was dreading doing a doubler, but it had to be done, so I carried on till the early hours of morning in which I had to walk home to Nantyglo, anyway I stopped as long as I could so my butty and I decided to call it a day, against the wishes of the night deputy, so I did not report for work the next morning. I don't know whether it was an act of God on my behalf because on June 28th 1960 in the morning that our disaster struck the explosion occurred at 10.30 or thereabouts that morning, most of my butties died and to think earlier I was working on that face 0-18 W district, just before it happened and that was June 28th, so by not turning up for work at my usual time I would have been one of the unfortunate ones so how lucky for me that those bunches were left on that face, my butties included Wally Davies the borer and little did I know that would be my last turnover on that fated morning, as I walked home from that doubler I could not foresee what

Blue Scars

would happen in the few hours I had just left the face, if I had carried on by the request of the deputy I would not be trying to write my biography and was the starting of my nightmares that will be with me the rest of my life, and today when writing this page I am not going to enter into graphic detail of what happened in that fateful day, only of my experience to which I took part in during rescue operations that was part of my duty. When the meeting took place at the canteen asking for volunteers in quite a few took part including myself, but before that I did not know of the disaster, it was brought to my attention while having breakfast when there were a group of neighbours standing on my steps asking my wife was Chopper at home, she said yes having breakfast, it was then I could see by her face something had happened because when she told me what had happened I was too overcome to say anything, knowing I had only left there about four or five hours previous, after coming to myself I decided to go back to the colliery and met I believe the workforce plus neighbours, the meeting at the canteen was very quiet but the chairman of lodge asked for order and asked could he asking for volunteers for a rescue party to which there was no shortage of and when I asked what part of the pit was involved and was told where it was, which in my own mind I knew somehow, that it was the place I had not left a few hours ago, it was my face O-18 that had taken the brunt of it and I automatically wondered what the position was, our place was well e4ntilated but we all knew that the presence of gas was around that area and was associated with our area, anyway I will not go into that but when asked would I be in charge of a party of six I accepted and was issued with a check for the lamp room and one for pit top, the brass one was around my neck in case of unforeseen circumstances but one of my party gave me concern, while we were halfway in, how he got there I don't know but it appeared that in the confusion and in the goodness of his heart he had joined the volunteers, fair play for him, but half way in he must have had an attack of claustrophobia being in a confined space and I can assure you that it was not the best of places to be in at normal times and he wanted to go back out. I could see that he was agitated but if he went I would be one man short of a stretcher party so when a team came with the body I asked for him to out with them while one of the other team took his place, otherwise I would be a man short and so it worked out that way, four on the stretchers and the other two there to relieve, but it was only till later we found that the man was off the Wonderloaf bakery van so that's how the baker boy got mixed up with the team and was not insured by NCB but fairplay for him he tried anyway, that was the trouble with the volunteers, in a chaos a lot of things are overlooked but with the best of intentions but the chap referred to did have the guts to try in a hour of need and that's all I am going to say about an episode that develope4d in our rescue operations because it will awaken something that I would rather forget, but not my friends that lost their lives but I had to write about it because it's part of my life history, but I will carry on with my story in a lighter vein to soften those traumatic moments and experiences that have passed through my mind, at the time of writing, especially what I saw and felt after I left my overtime shift at the 0-18 conveyor the night before and what a lucky escape I had by not turning up for work the following morning, by being worn out by those last two shifts that I worked on that ill fated conveyor but speaking for myself I have come through some comparatively narrow escapes myself but as the favourite words of politicians say lessons can be learned and

Blue Scars

especially learnt from June 28th 1960 at Six Bells. But after that safety awareness were foremost in everybody's mind, that is why I was encouraged to do those safety sketches.

So now I move to a different scenario and into reality because of the technology of power loading monsters of the deep, that has taken away the developments of small belts of the development era so I was out of a job so to speak and was back to square one again to supplying supplies to those that needed them and sometimes delivery items that were harder to transport, like the transporting of a segment of a disk cutting machine that skimmed the face of those power loading conveyors, the disk itself was about four or five feet in diameter and was made up of segments and a part of the disk segment had to be loaded by crane at the surface so what hopes did we have of the trolley carrying that part of the disk was derailed, I dreaded to think, all we had for lifting gear was a Sylvester, the segment reminded me of a inverted ice cream cone, but a bit more heavier than that I can assure you, so down the pit it came and was shunted to the supply journey and there to be taken into the back of the work, to its destination, the power loading conveyor, through the twisting narrow parts of the roadway where it was needed to be turned on its side in some parts of the roadway that had low headroom and did not cater for such a monstrosity, but it was a test of the skills of the supply men, the gaffer in charge and myself to manipulate it to its destination, but we got over that and most of the difficulties and disagreements which we all expected to happen by frustration and a few heated arguments but the main problem was the unloading of the damn thing, but the only way out was the Sylvester was attached to the ring arches which was dangerous and not very safety wise, finally raised the segment above the drams edge and roll it back underneath and lowered the segment to the floor but we encountered our usual enemy which was water again and I remember the other incident we had earlier with a heavy roller but this was different because two men on each side had to balance the segment when they tried to bowl it quietly through cable tines and rails that you could not see owing to the fact that the movement of the feet had muddied the water and if that heavy piece of machinery had toppled over it would have trapped the men with perhaps fatal consequences and the rest of us would be unable to lift the machinery off them, so the segment was inched forward very slowly but after an hour or more succeeded on delivering it to where it was require, but after that the face continued till that was finally abandoned owing to water and to men like myself would not relish another trip down there in those circumstances but every day in the pit was a challenge.

Sometimes these roadways when they travelled longer distances needed extra ventilation and my one district needed that extra ventilation and a road adjoining the main airway had to be made, it had to be made and driven at an angle to meet another roadway, but funny enough, it had driven into a roadway that was familiar to me being an old carrier belt road which carried the coal from the coalface by the name of 0-6. It was a conveyor that I used to turnover years ago but now it was a linkage for our return airway and when it was completed things seemed to take a sinister turn, I don't know if it was a coincidence or not but that roadway had been blocked off after the explosion because if debris from the disaster and other materials had been buried there and not to be uncovered by the blocking the mouth of the road, but had to be opened up for the continued air passage,

Blue Scars

but later on a few of us felt a strangeness in the atmosphere and felt that there was a presence of something that no-one could account for (and I mean this) and sometime later on were a bit reluctant to travel the road, including myself, although previous to this I had worked many hours on my own repairing the rail track and other jobs that needed doing and thought nothing of these rumours that were circulating around the area, but on occasions you felt uneasy but could not pinpoint the source of this uneasiness but it was always there. I give you an example, on the one occasion when we were short for supply men and I had a new butty for that day and he was complaining from the start of the day that he was not getting the same rate of pay that I was getting, but that was up to the discretion of the overman, not me, he continued this conversation all the way to the curve in the road in which we had to go in front of the journey to clear the track if need be, so halfway down the track he still protested but on seeing a haze of light in a disused roadway where men used to have five minutes whiff out of the way of the wind so I told my new butty to shut up and not let everybody know our businesses, so we came close to that roadway and finally entered it, but there was not anyone there, we looked at each other and I said to my butty, did you see that haze of light that was in this roadway, hoe looked at me and said 'Well aye of course I bloody did', well I said whoever was here and if they were they could not go through that door behind them because it had been sealed off by nails driven into the post at the side of the door, so I said it must have been the reflection of our lights on the oil or grease of a nearby haulage engine, to which in my own mind know that this was not the case because of the engine being too far back but anyway it was a bit unnerving to say the least, so we changed our journey and transferred

the rope to an empty journey of drams to be taken back to
the pit, I said to my butty I will go up and sheave the rope
around the curve and you will have to put the bar hook on
the last dram because if there was a runaway journey the bar
hook would throw the last dram off the road, so I made my
way up, but on approaching the curve my butty was behind
me, I said 'I told you to put the bar hook on' he said 'Bugger
you I'll come up with you, let the journey look after itself',
he was a bit bewildered so I took the chance of bringing the
journey to the pit without a bar hook and that was that, but
I still believed there was an explanation to that reflection
and when my butty recounted the incident I was the recipient
of a few jokes until one day a reliable deputy was standing
by our haulage waiting for us to drop our journey of supplies
into the parting (double rail) the engine driver put his signal
bell on for us to start but the bell gave a consistent ringing
so that meant that somewhere in the journey road the twin
wires of signalling had crossed one another and they had to
be parted to stop that continual ringing of the bell, so the
deputy suggested that he would continue his rounds and
while on his way he would find out what was the reason for
the ringing, so off he went, we all waited for the ringing to
stop, when finally it did we went on with the job. But these
occurrences of crossing wires were frequent by fallen pieces
of stone or by a dislodged piece of timber from the ring
sections, we proceeding with the journey of supplies and
reached our destination (my regular butty now though) and
started to unload the supplies when the deputy called me
further up the journey road, he was a big puzzled and
inclined to be a bit agitated to a certain degree, he was
normally a sensible person and well liked and respected and
always quietly spoken gentleman by the name of Don Meek
from Llanhilleth but anyway after calling me up he said to

Blue Scars

me, don't repeat what I am telling you because if word got around it would question my sense of responsibility to I told him that I wouldn't repeat it, so he continues his experience while looking for those crossed wires, while doing so out of the corner of his eye something resembling these balls of matter you usually see on a Western film, the things they call tumbleweeds blowing up the roadway in my direction like a misty ball of wool. I stood in the side and it passed me and disappeared into the stonework of the roadway, I examined the area but there was nothing there, I said I believed him and now I said someone may believe me but I won't be able to repeat what you said Don, but it will be hard to keep silent but I was glad I was not alone in my belief, the forces were at work and not in my imagination that something emanated from some supernatural sources but talking to Professor Harvey that was in Aberystwyth University that many pits can testify to the phenomenon to visionary encounters from the same phenomenon and now manifests itself from one of my own paintings, on the anniversary of our disaster which happened at Six Bells in 1960mbut the painting in question was painted in practically the same time but was in 1985, twenty five years afterwards and showed miners images trapped in the painting which has baffled TV critics and national newspapers including a talk with Roy Noble in his talk show at Cardiff Studios but it is the only painting out of approximately 400 that I have done to have that vision. But to get back to my story that I will not deviate from of other incidents that would be substantiated by officials and working men alike, anyway while driving the haulage in that same area and I always carried my food around with us to different jobs that required us to stop longer than normal, my food was always carried in an old army haversack to stop the mice getting at it, the back wall of the engine house was built out of breeze blocks and the breeze blocks had two square holes in it and to be doubly sure I squeezed the haversack into the hole and the deputy hung his food on a length of wire from the ring arches, so while my butty had gone ahead in front of my journey to clean any debris that may derail us and to make sure that there were no one travelling the road and I awaited my signal to operate the engine and release the journey, my butty was well out of sight by now and I was hoping that he would hurry up because I was getting a little uneasy by now in the silence, anyway the signal came to drop the journey down the gradient in front of the engine so off I started and watching the drum revolving in case of derailment at such a speed until it reached its destination but having pegged the brake down there was no haversack by my feet, and the first thing that I thought was someone was playing a joke on me because I would not have heard them because of the sound of the drum revolving, so after taking my lamp off its hook I said "come on you buggers, you have had me", but there was no answer, so I walked up the roadway at the back of the engine peering in the manholes (refuge holes) and they could not conceal themselves anywhere else and then I got a bit panicky, left my haversack on the floor where it was, I ran like hell down the journey road as fast as I could because I wanted to be by my butty's company. They wanted to know why I was in an agitated state. My butty and a few of the men started to laugh when I told them "come on and unload these supplies that will take your bloody mind off things", I thought to myself what's the good of telling them anyway but after the unloading of supplies I said "well you can please yourselves but I am not touching that engine till one of you come out with me". So my butty said "alright we will sheave the rope around the turn and I will come and walk out with you" and that's what he did, my haversack lay where I had left it and I would not drive that engine again unless I was accompanied by one of the boys, but what I couldn't understand was that I had worked on my own for hours doing different jobs so I didn't know what to make of it I'm sure, but all I know after the linkage of the rope where the bits and pieces from that explosion were buried, it seemed to have changed the whole atmosphere of that district, but in the event of the closing of O-13 and O-17 the parts of the area was abandoned except for salvage operations of pipes and essential machinery and that took a decent time and the only time I visited there was when I was in the company of the salvage man and now the pit has closed and all the pit head gear has gone, I often wonder what is happening in that water filled graveyard of memories and I have often thought that whatever happened below, no one will ever know, but there are a lot of strange happenings and who will believe you but I and others have experienced that feeling of unrest that lie below. Talking to a friend of mine by the name of Mike Gaylord of Vincent Avenue who were talking about these phenomena that has happened and he says that he himself has had some strange feelings of what happened to him while working at a black vein LB8 conveyor cutting bolts off the girder work with a chat (heavy duty chisel) and sledge hammer, the pieces him and his friend had cut off were thrown into a heap to be loaded into the drams but on one occasion he was throwing a piece of cut off sections onto a pile of irons, he was sweating but a sort of cold sweat, he said he could hear the men working and talking on either side of him but when throwing the sections of iron onto the pile could not hear the sound of metal against metal and the cold sweat continued for a few moments he said the experience felt like the old saying of like someone was walking over his grave but this feeling he had was not of Six Bells, but my old pit of Beynon's Colliery, so these happenings are not only confined to one colliery and could have happened at any colliery in the coal fields.

I was offered a job as a track layer later on of which I had not alternative but to accept but it kept me in employment and I accepted the challenge of my responsibility and as it turned out I became well adapted to the job and kept me from being pushed about from pillar to post. I was a bit nervous at the beginning when I was in the vicinity of the pit bottom with its complicated arrangement of tracks but it wasn't called for any of that because the track was mostly embedded into the brickwork and were properly secured and in good nick thank god, but my main job really was to maintain the journey roads and supply road to which I was authorised to do, a far cry from the dodging of travelling journeys while accompanied by an official, but now I was more or less seasoned to the job and had a few hiccups as were expected but it was part of gaining experiences of the job. I used to be aggravated by the traffic boss on a few occasions, a thing I could do without, but his job was to keep things moving I suppose but things got a bit complicated at one time and a supply journey came in with a few trolleys of broken up rail tracks. I wondered what they were for and was told I had to assemble them into a double rail track called a parting that would hold an equal supply of drams on each track, full ones and empties on the different tracks, they were taken into a roadway in the airways on a single track. It was a decent size area inside one of the ventilating doors and had to be re-assembled there. Oh hell I thought I have never attempted that before, so after the unloading of this pile of rails and sleepers by the supply men the journey was taken away and I was left with a pile of rails and sleepers to which I had no idea what to do with. I thought to myself Chop you have bloody had it, but they didn't need this parting for quite a few weeks now and I thought it was alright for them on the surface because they had plenty of room to work but I had only the area to assemble it in a confined space, plus I didn't have a clue anyway, but I had a few weeks to fathom it out and I would have to have help in any case to manipulate the heavier items, but luck would have it the parting had to be laid between the ventilating doors and men from the loading head came inside the doors to have their food, out of the cold air, so one day one of the oldsters shouted at me that I had placed the crossing piece in the wrong place and that brought other men into the argument, one was saying this and another was saying that so they started to draw a sketch on a piece of rusty tin. I didn't say anything but listened so after they all agreed with each other the old man said to me "there you wass (boy) have you got the idea now?" I said I was grateful for their ideas and he said "it was something for us to argue about anyway, next time we come to grub I will give you a hand and I expect the rest of them will join in too, so don't worry Chop but if you have any spare bacca that would help I suppose, what do you say?" "right o" I said, and under my breath I thought you crafty old bugger. The next day I brought their bacca, it was well worth it too. The parting was laid in a fashion except for the bolting and other knick knacks but it was all under control and I found out that there was more to track laying than I thought, but a few weeks later it was put to the test and after a minor adjustment the operation was put on a serviceable basis and I was proud of my little accomplishment, and whenever I was around that part I always had a bit of bacca to spare.

Chopper Davies

Looking back over the years and thinking over hard ships that I never knew and families used to suffer and as a kid I would not know of, but I was never without food and everyone in our village depended on our school hand outs such as boots at certain times of the year, but those hard times do not compare with what is happening today in which civility friendship and respect are all gone to be replaced by hooligans, druggies and vandalism. I am not saying that these phases of this sort did not happen in my day but not to the extent of what is happening today and the harshness of them days were softened to a certain extent by local festivities that were held by the community. Things like carnivals where all kinds of vehicles were used, mostly coal lorries decked out by buntings of various colours and costumes worn by contestants to denote to what they represented, then a cortege of characters that made the procession more colourful and symbolising and typifying the characters that they represented, but there were scary moments as well when a character naked but for a loincloth leapt out on a crowded pavement but it was all great fun, except for a few screams by some frightened women and kids and it was all done in devilment and good spirit of the carnival in which invariably ended up in some sports ground or fields nearby or anywhere there was available to continue events for the day, in my case and some others to an adjacent field nearby and that was Dan's field and that was adjacent to our local hospital. That field is now a building site called Cae-glas but there was nothing spared to break the continuity of the day's events and all helped by the generosity of the local shopkeepers. Their usual packets of sweets and popcorns and crisps but the day would not be content without the old favourite one or two ice cream carts (not vans) that absorbed most of my pennies, the

Blue Scars

flavourings of different colourings that mostly ended up on the front of my new jersey and another event was always welcomed was the school walks and the same procedure as the carnivals, but we would have our best clothes on then, the same sweets and the same old ice cream carts and some scrambling of thrown sweets into the air in Dan's field of which I stood no chance of getting by being knocked out of the way by the bigger boys, but had to be content by scrounging in the long grass. We were always accompanied by the dignitaries of what chapel you went to and afterwards a rush to the vestries of the chapel to a feast of pastries and bottle of pop, till you could eat no more, then off home to change out of your best clothes. The only time I had a new jersey and tie and a new school cap with a nice new badge on the front of it, then if you were lucky you would be taken up the fair in Brynmawr by your father and mother. You can smell the burnt oil and grease that emanated from the steam engines and I watched the pulleys and belts that whirled by the running of those powerful engines with their brass fittings flashing in the lights of the fair ground. I used to watch the performance of the dodgems careering around on the steel floor and the electrics flashing overhead from the netting wire overhead by the power that controlled those little cars bumping into each other amid the screams of delight from the occupants of the car but what amazed men was the agility and nimbleness of men collecting the fares from the passengers without falling under the wheels and it was a feat that only the fair men could muster, then we would wander around with my eyes on the ground in case someone had dropped a penny. The only thing that I am sorry about now is my father won a goldfish and I carried that fish around all day, although I was excited as all kids would be with a goldfish and could not get home quick enough to put him in a jam jar but thinking about that poor fish in a plastic bag being carried around the fair, it was not fair to see that poor bugger trying to swim in a plastic bag full of luke warm water, it was not very nice at all and I think in my own mind that would be one of the things I would certainly put a stop to, but after walking around all night I was buggered up after a few rides on the Noah's Ark and a few other rides. I was ready to go home and up to gwelly (stairs) and dream of the day's events, finally dropped off to sleep, even too tired to read the comics, the Dandy, Beano and the Wizard was all forgotten but the coming of another day would be a repetition of the first day, doing errands to get a couple of pence to go to the matinee at the Cosy cinema and market hall Brynmawr and if any coppers were left might permit to buy a crafty fag called crayals and a bag of tiger nuts to which I shared with the boys and swapped puffs between us out of our crafty fag and concentrated on our favourite actors like Gene Autrey in the western features, with a mixture of Laurel and Hardy thrown in stop and see next week attractions and if you did not like them you could always go up to the Cosy cinema but sadly today the Cosy is demolished to make way for flats and a shop, a sad day in our time, and replaced by steamed up drunks loafing around with cheap cans of lager from supermarkets and off licences, if this is the third generation what is the fourth going to be like? I can't bear to think.

But we did have our sad moments in those days when I used to stand on the top of the tip near the old Mount Pleasant public house on top of Garn road and watch the long line of men in bowler hats in a funeral procession, the dignitaries with their Homburg hats, these funerals were not only attended by the family but I think the whole village were there to pay their respects, a far cry from today, the respect is the same, but the cortege of men and women have whittled down owing to the transportation of cars, but I wonder would I have changed but no one knows your future but I am glad that I didn't change because I would not have met a wonderful wife because we both had a lot in common and we came from the same street Vincent Avenue. I lived in number 3 and my wife lived in 25, if I had not met her then I don't know how I would have fared, she gave me the meaning of life and love and was the mainstay of us and kept me to a more stable life, I may not have had otherwise, so I was one of the lucky ones but now even that has changed, after losing her, but she would have wished me to carry on with my work as an artist and I will abide by her wish to try and get that spark started into a flame again, but after being married for very near sixty years and losing her very nearly finished my ambition, but I am fighting it with the help of my two children.

Getting back to these strange occurrences that was dictating my way of thinking and what was happening and could not pin point the origin of its source, but more so at night, by you could put that down by the cooling of the earth surface as you went into the early hours of morning but the unnatural of a night shift was half the cause of the eeriness that happens between the hours of two and three o'clock in the morning but to a miner on that lonely shift every sound was amplified by the absence of running journeys and an army of men working. In the silence your imagination takes over specially when one is working by himself and I can tell you I am not the bravest of men after my previous encounters as many men can testify to, but when mother nature puts her foot down there was the understandable groaning of those wooden supports like if mother nature protesting of the raping of her body from the predators who rely on that quest for those black diamonds, that makes the valley survive, but the miners have given something back too, in lives of many miners, but that's the risk and whatever precautions you take you cannot avoid some sort of injury and leave behind many a blue scar.

But to get back to track laying, which became more than complicated like for instance when I was sent to install safety appliances, safety latches, similar to some on main lines and in shunting yards where you have to throw over weighted levers to direct the flow of traffic, but these were to do the opposite, should a runaway journey be involved or loose drams breaking away, but after drawing a few diagrams on the side of a rusty dram I finally mastered the problems and I thought after what was I worrying about, but I was thinking of the responsibility that was on my shoulders, should my appliance fail and something go wrong to endanger someone's life including my own, these little things you may forget was the cause of so many prolific accidents, so you have to have confidence in your work and yourself, there were times after doing and finishing a job I would travel the transport planes to check for loose bolts or fish plates that had loosened, owing to the hammering of those heavy journeys of coal. I sometimes accompanied the deputy on one of his rounds of inspection, by doing so I checked disused rail tracks in case of some of these fitments would be useful should they be required in other parts of the roadways, but I believed that sometimes it was to cover some of his nervousness, in some parts of the airways, but only in that one area, to which I won't mention, but it worked both ways to cover my nervousness when collecting parts of rail track, so all in all it was in the interest of both of us for the company of each other, but it was interesting to watch him at work testing for that elusive presence of gas which is always present in deep mines and my place of work was no exception, but the standard of safety in our colliery was unsurpassed, with all credit given to our officials.

But when I was not too demanding in my job as tracklayer I was asked on a few occasions to supervise and give advice to new entrants into belt turning, such as how to correct a doubling of the belt at the loading heads and tension ends, because I had no forgotten the tricks of the trade I had learned over the years of experience, but I could not compete with the younger element in heavy work, as you grow older and the older you get the time seems to pass quickly. I found out to my sorrow when helping out, I was trying to shove over a tension box at the end of the conveyor but the chain attached to the box was buried under the rubble of the roadway un-noticed by me, and my back was pressed against the wall of a road side pack and I tried to shove the tension box over into the new line when I felt something give way at the bottom of my back and a warmness started to spread down my legs. Bloody hell I thought I've busted my back so I managed to crawl out into the road way with some difficulty and tried to stand up, I did so with some effort and I knew that when the circulation came back into my legs I would feel the extent of the pain and I started to walk and with some difficulty reached pit bottom, after which I should have been carried out, perhaps I may have caused more damage but it put me out of work for quite a few weeks but when I started back I knew my career as turnover man was ended. They say that hard work never hurt anyone but they will have a job to convince me on that, but I did continue my work as a track layer but with more caution this time and even that work demanded a lot of lifting, but I engineered it so that there was someone in the vicinity at the time I needed a lift with anything that needed an extra pair of hands and I was in a better position than some of the men at the colliery, so I had better stop bloody grumbling and get on with the job I was supposed to be doing. So when I had a quiet day I would brush up with experimenting with the oil and grease and the usual piece of chalk and get reprimanded again by management, but it was worth it but even some of the officials would find time to chuckle at some of these embellishments that were added to my drawings and that gave me an added boost to continue, so one day a haulage driver had gone absent and I was asked to replace him till he returned. I did not know for how long but I was told to take over the role of the engine driver and while I was there I was attracted to a large expanse of brickwork and the white lining crew had done a good job of whitening that wall and I was in a dilemma trying to resist the temptation to do something about it, because that wall was tantalising me and I wondered would it be worth the risk to have a go at it, but the love of art and the easy access to materials that was needed overruled my guilty conscience and I decided to disregard the consequences that might be, but I wondered how long the driver would be absent because once I started I would have to finish the project I had in mind because half finished it would like I was defacing the wall, so I had to make sure whether I could complete what I had in mind and I knew the manager would not be around for a few days and that would give me time to disappear. So I started after telling the boys to let me know if the manager was coming on his rounds of inspection and I was also depending on the good nature of the manager, a Mr. Arthur Lewis who was a good manager and my favourite out of all the manager I have worked under during my stay at Six Bells, but I took a chance and it was like if luck was with me because his visit to our district was delayed owing to trouble at another part of the pit, so it gave me ample time to complete my work as long as the other driver stayed away a bit longer, but on my third day I had nearly completed my work the driver had not started back to work and the manager was expected but I did not have much to do anyway only to sprinkle to sootiness from underneath the rollers and let it soak in the oil and grease and highlight some of the painting with chalk, it was a portrait of a cowboy, eight feet in height to which in my mind was a masterpiece and I was proud of it, not that I had completed it, but it stood out starkly in black and white. It was stunning and it took the attention of everyone passing by and I was congratulated by them, but I wonder would it extend to Mr. Lewis's point of view. I would have to wait and see. What I was concerned about was where in hell was that driver so I could get away from there in a hurry but he never turned up so I was stuck with it, so came that day when the manager was expected and they warned me he was on his way into the district, I thought to myself bugger it, I was trapped there and could not very well avoid him because he had to pass the engine and I hoped he wouldn't want to talk to the driver. I bet the boys would be laughing their guts out because I would be faced with a predicament to what the manager's reaction would be to that awesome looking cowboy, I could see by the reflection of their spot lamps as they slowly approached, I did not know that he was accompanied by the inspector of mines, they were coming nearer and nearer, suddenly the lights had stopped and I could hear the murmur of voices. I don't know if it was the manager's voice or the inspectors, but I could hear him say "well, well, well, what do you think of it?" but I noticed that there was not any sign of a reprimand in the voice but a surprised sort of humour and I thought I detected a bit of a chuckle out of one of them and I hoped that the chuckle was that of the manager's and would soften the reprimand that was surely to follow, then they were beside the haulage so I go busy the other side of the haulage and was busy oiling that side but the manager spoke to me and asked "how is the coal coming driver?" and I had to answer didn't I, and that was my downfall, he shone the light on my face and asked me where was the driver, I explained I had to replace him when he was absent, the light went off my face and he spoke to the inspector "this is the chap that is drawing about the place and defacing the wall after the nightshift try to make it more illuminating for the roads". The inspector did not answer for a moment and when he did I had a surprise at what he did say, "well it does not look too bad and it's an unusual way of defacing walls I admit" with a bit of a smile, but the manager spoke to me and said "this is the last time Johnnie, and it's got to stop understand?". It was natural he had to show some sort of authority even if he did not mean it, I said I was sorry and I know I did not mean it either so off they went but I could hear the inspector chuckling and I was wondering what the manager was thinking also, but I don't think that there was any real reprimand coming my way and the rest of the day went bloody lovely now that the pressure had gone and I heard after he was supposed to have said "I should find something more suited than white washed walls" which I did in the future and develop a sort of photographic memory that I was to use in my detail of mining pictures because mining art became an obsession for me to which I took advantage of during the thirty five years of drawing and painting it came to me to collect all my resources of men and machinery to compile approx four hundred paintings to illustrate my work, because even one of them can speak a thousand words and I have missed through the countless hours of social life by doing so, to try obtain some sort of status as an artist. I am not anyway of the opinion that I may make a great artist but in my humble way try to pay tribute to the millions of minders that lost their lives in their pursuance of coal and also to our pioneers of that past era and to preserve our rightful heritage that future generations will not understand and with the help of the likes of museums and galleries and myself will keep those mining years alive. We all know that we cannot live in the past but the artefacts that surround one show that I still do and still give me that thrill of being surrounded by the memories of the camaraderie that once existed among us, but there are vast areas of untouched energy that still exist underneath our very feet but climate changes have put a stop to that but there are ways of harnesses that energy if technology was put to good us to reduce the carbon but anyway that was far beyond my way of thinking, all I am concerned about at the present is to be classed as a caretaker of memories and a trustee of the past. Like I have stated I am not a Picasso or a Rembrandt but in a way I have a following in my work so I have been told by some critics and I have the same passion in my work as any of them and show the strongest of feelings to what I believe in and that's the most important and must have captivated the attention of the coal board and management alike who commissioned me to exhibit my work throughout the South Wales coalfields. It was better than drawing on rusty drams and white washed walls but if it wasn't for that I would not have been noticed in a creative world of art as many schools can testify to and I am proud of what I have accomplished over thirty five years, but age waits for no one and I am not a young man anymore and you cannot choose what you would like to be remembered for and I don't know if this book will ever be published or if any of my paintings will survive although they are spread over different countries like America, Australia, Britain and New Zealand but at home who knows? But can only be found in ones heart but my story sets a precedence for any future artist that undertakes the role of a mining artist and remain with me in the archives of the coal board and TV alike but in schools they were more appreciated not only by the children in their thirst for history but by the parents of the children who collected them from these schools and that is when you yourself get the greatest thrill of all, with the explanation of what was happening in the dark labyrinth of tunnels under their very feet.

One day I was called to the office by my favourite manager of Six Bells Colliery, Mr. Arthur Lewis and I wondered what I done this time because you don't know what to expect from day to day but to my surprise it was not a reprimand for which I was expecting to have but I was asked by Mr. Lewis could I like to take part in an exhibition of mining art. He explained what he would like me to do and the reason for asking me there was a competition of safety held by about two hundred artists in the coal field. It was not the usual exhibiting that I was used to but for artists to show their paintings on safety values and he would like me to enter as an individual and submit two or three of my own from Six Bells Colliery and would put Six Bells on the map so to speak. I was excited and delighted by this and accepted because fair play Mr. Lewis always encouraged me in these sort of things, I thought to myself, I don't have much of a chance among that lot of adjudicators but he assured me

Blue Scars

that everything would be above board and not only that but he was on the panel of judges. The prize was not a lot I don't think but it was not the money I was interested in but the honour of Mr. Lewis asking me to participate in this exhibition and I felt some pride in that, so like I said I submitted about two or three could be more but don't remember so they were handed to my manager who was very pleased who said "don't get disappointed if we don't win or don't be disappointed with the results because there are hundreds of entries". I was not concerned how many were entered but having the honour of representing Six Bells Colliery and I was confident enough about my own ability to stand up to the best of them. I didn't know how long this exhibition was going to be but it was not coming soon enough to suit me. I was pleased to know that as I was the only one to be asked to represent Six Bells so judgement day cane and went and I was called to the office and asked to sit down and gave one look on his face but told me not to be disappointed at what he was going to say and he himself was not disappointed with my efforts. "well Johnnie I will tell you what happened, we were well ahead of other entrants and it came down to two or three of which we were among them but by a narrow decision and it was decided that the prize was awarded to one of the inspector's sons and Six Bells was second but it was very close so you did not ought to feel so bad". I said then I had to because I could not control myself at that point not because of letting Mr. Lewis down by the decision, I said then "what do you expect he was the inspector's son wasn't he" he could see by the way I said it he did not say anything of what I implied but he did give me a small smile knowing what I inferred to but it did give a feeling of small comfort to know that I had made an impression from about hundred or more artists. I didn't even

Chopper Davies

ask who won it, but none of my entries were returned to me and often wondered that maybe the professional artists would pick the bones of all the artists' entries for their own use. I thanked him for having confidence in me and proud of him doing so, and that's the truth and that ended another chapter in my life and will soon be forgotten but not by me and the story I tell is also the truth too.

This part is taken by another manager who succeeded my old friend Mr. Arthur Lewis, his name was Mr. Peerce who I was not particular attracted to nor him to me either. A different man altogether than his predecessor but that's beside the point anyway. I was making scenery for our annual exhibition at Roseheyworth club in Abertillery prior to our exhibition at the memo club in Newbridge, it was between two teams of colliery workers, Six Bells and Marine Colliery Cwm Ebbw Vale. I had completed the backdrops for our scenery and I was satisfied, Mr Preece had seen it also unknown to me until a few weeks after, before the two teams were to meet. I was working at my job as tracklayer, just potching about, the coal was stopped for the descending of the cage that carried men for that moment and who stepped off the cage was the manager Mr Preece and another man. I could not see from my distance away, they both had spotlights and oil lamps, so I busied myself with tightening bolts on the track that did not need tightening anyway. I was bent down with my back towards them but moved over to allow them to pass but they did not pass me because they shone their lamps directly at me and I thought I hope they would not examine me too close because my helmet during the years had seemed to have screwed around and my headlamp was not shining in front of me but sort of on the side a bit and furthermore than I didn't have any sort of safety boots

Blue Scars

on, boots yes, but with no steel toe caps so I managed to bury the caps in the dust at the side of the track. I thought to myself I am a clever bugger to paint safety scenes which I didn't put into practice but the manager was not interested in my attire but said to his companion 'This is the chap I was telling you about the Safety Exhibitions', so when he addressed me I stood up and he said, this is the chap that is doing the scenery Mr Cartwright. I knew Mr Cartwright from a few occasions but not to speak to but the manager continued and spoke to me 'I am satisfied with what you have done but I would like something leading up to it, like an older man telling the younger man about the safety aspects of mining, but this conversation must take place in the atmosphere of a bar in a pub where this conversation would take place and have all the effects of barroom talk'. I told him that there would not be enough time as the Exhibition would be taking place in a week's time and I would have to paint a backdrop to cover the other scene by ten fee by eight feet because it would involve the painting of the array of bottles, optics and cigarettes, plus the bar attendant and it would have to be perspective painting of the floor leading to the bar scene. He then told me to take as much time as I needed to complete the scene, he meant in so many words, that I may have to take time off work to do so, fairplay to him, so I worked over the weekend (in my own time) in the old lamp room where I had plenty of quiet and space and sometimes at home, I didn't work the first Friday, the starting of the picture I had in mind, I worked the weekend and missed my couple of pints with the boys. I worked all day Monday following, missing work, so more or less I had two days off from work but was encouraged by a member of the cast that was to act in the play by the name of Don Taylor of Abertillery, whose wife made sandwiches and a

Chopper Davies

flask of tea and delivered them to the lamp room, where I was finishing the scene. I finished my work of painting the backdrop that provided the scenario for the cast of actors and then reported to the manager that it was completed. It was two days before the competition, as I entered the office, the manager's clerk said that the manager was in conference with the reps of the mining industry, but I took no heed of him and knocked on the door with my rolled up scenery and was told to enter, the manager looked at me for interrupting and when I explained the situation and the reason for the interruption all he said was 'Have you completed what was asked of you'? I said 'Yes', he said 'Go to the lamp man and tell him to book you for Friday and Monday and if he wants confirmation that I asked for was to phone him' and that 's what I did, I then proceeded to the carpenters shop for the work to be glued on hardboard and that was that, but I did not wait to see what it looked like all in one piece, but was confident of the outlook that it would make, but what I omitted to say was, I had painted the door of the lounge adjoining the bar a black cat outlined against the frosting of the door panel with its tail erected, which I added for extra luck, well at the end of the show (in which we won) I went to the bar to get a pint, I hears an official from the competing side say 'I know what I would like to do with that bloody cat', he said it in no uncertain terms too, but we won. One of the adjudicators was Mrs Peggy Witcombe of the Blackwood drama schools associated with Blackwood theatre plus inspectors and officials of both collieries, but in respect for the team of Marine Colliery, they put a marvellous show of a short firing team and I think we were lucky (or was it that black cat). For weeks after that I used to come up the pit at twelve o'clock to a cup of tea and a fag to watch the budding actors perform for the next competition at the

Blue Scars

Memo Club at Newbridge, which also proved a success against North Celyn Colliery, but yet another competition was held at the Manor Suite Porthcawl, where I had to go beforehand to measure the stage of our scenery which were different but I will not go into that as it is too lengthy, but all this was contributed to mine safety, otherwise there would be no point in holding these competitions but it came now to the semi finals which was held at the Pavilion Porthcawl and that was very near the pinnacle of our success, but not only for me as an artist but I had a double role because I had to be behind the scenes as a prompter in case one of the boys failed their lines, I had made small holes in the scenery to allow me to watch and prompt should they fail in their lines but the audience could not see me or hear me if that time came, but the audience had increased over the years to two hundred or more owing to the following of the competing buts, but unfortunately things had to come to an end sometime and that happened at Porthcawl when we lost by half a point in the semi finals that would have taken us to Blackpool for the finals, but we had to smother our disappointment like the other competitors that failed to us, but I was proud to be among miners that contributed to the shows, like the miners from Ammanford and parts of Yorkshire that brought their followers and like us our wives, so that was the end of our early trips up the pit and everything was back to normal again.

Back to my reminisces of Beynon Colliery, Blaina, that was the start of my doodling on rusty drams. I was a tip end haulage driver and it brought back memories of some of my friends of haulage work, there was Bryn Collins, Tom Howells (Gunner), Jackie Best, Joe Monk and quite a few that never worked in my district of top coal, some worked

Chopper Davies

in a lower horizon, such as Hollies and the Meadow vein but we were all boys of about the same age and a few have passed away now, but we were all of or part of the workforce but my interest was doodling on rusty drams. Some of these drams were used for debris, such as much and broken pieces of timber and so forth that were taken to the much tip to be scoured for coal by the muck pickers or coal that may have eluded the screens, but I knew some of these coal pickers so I had an idea, that why didn't I tell one of them to watch out for a dram with the drawing of a cowboy on the front of it, so when a dram half filled with coal was handy at the time, came in handy one day and I stuffed all sorts of debris on top of the coal, to which would be sent up the tip with other drams. I drew a good picture of a cowboy in chalk on the front of it, it worked fine but I had forgotten on other occasions to tell the coal picker that some of these drams with cowboys on fronts were in a circuit that travelled around the districts, by that time cowboys were on a good many drams by now, but every time a dram came up the tip he would claim it but I did not have chance of letting a dram of half filled go all the time so he got fed up of looking for that elusive cowboy, but I was afraid to tell him that I had drawn on most of the drams, so I kept out of his way for a long time till it wore off, because I was afraid of having one across my chops, so I did not bother anymore of doing favours of that kind, that was the only time my cowboys let me down, but Beynons finished in 1975 and I expect most of paintings in oil and chalk are underwater now but they will take some removing by the water that had filled the pit, but that pit will hold some memories for me and others.

But time to pass us by and my back injury did not improve and I wondered again how long I could keep it up, but carried on for a few more years and on one occasion thought seriously of packing it up but I relied on my experience to get me out of most difficult situations and it held me in good stead. I knew that like others before me the dread of leaving the pits because I would not be in a position to be offered another job because of my age and like others I would be low in spirits and get despondent once I handed my lamp in for the last time and I know also that when I did that I would lose my identity and all the friendships I had created over the years would be lost and you would be on your own, only for special occasions and that would be getting less as the years passed by. Maybe I am a bit of a philosopher and devote myself to the moral principles that I study and investigate the wisdom of living down that dark world. After two thirds of your life had been spent there, I would miss the clanging and screeching of machinery, the smells that emanate from different parts of the pit and the intrusion of family life of my mates. I would miss the bantering and sometimes that odd quarrel that would end up more friendlier than when it started, I would miss the foul air and decaying timbers that emanate from those labyrinth of airways, all this would bring back nostalgic memories for me in my lifetime n the pits, but the position I would be in, is that I have a hobby of artistic qualities that would keep one on my toes mentally and alert, that was or is one good thing that as came out of it, because if you let yourself go to the dogs as the saying goes with no incentive to carry on, then you can call it a day, but even in the increase of technology today in mining, that would seal the fate of the coal industry, the strike of 84 proved that by given the Thatcher Government the opportunity to cause her mischief, had there been a scarcity of coal at that time she may have had second thoughts, but our boys in that magnificent struggle although futile will not bring it back but history will not forget the mines and their struggles.

One morning I was drawing in pencil trying to increase my ideas as usual of one of my paintings and a feeling of my arm getting heavier and a bit weary and a sort of numbness creeping down my arm and I thought to myself I am overworking my arm by the amount I was putting on it by my work and the factor creeping also in my mind was I was having a stroke, so for therapeutic reasons I kept on drawing outlines of what I had in mind for the painting and after a few minutes or more it more or less came back to sense of movement that I could continue but I completed my painting and that was that, so I went on with my ordinary chores around the house, but it made me think that all the paintings that I done, it was the only time that this feeling had come into my arm, months went by and my daughter Elaine at that time was working at a local garage as a petrol pump attendant, while doing other jobs pertaining to her job, while cleaning the flat above the garage and was dusting around was going to clean a certain picture of some artist, and her boss told her not to clean it as it cost a lot of money, so she mentioned that I did a lot of the same kind of work as that artist, such as mining, so was asked to bring it in and show him but she showed it to her friends that were the mechanics and they said to her, take it from here because they said they could see images of faces in the painting like ghosts, so when it was brought home I studied the painting for something and eluded me and there was no mistake at what the garage fitters saw, it was undoubtedly images were there distinctly, plus below the face further down the painting was a hand holding what seemed to be like the handle or grip of a blast pick, so I did no more but showed it

Blue Scars

around the clubs and pubs like I always did when I painted a picture of the pits, they also recognised the meaning of the hauntiness of that particular painting, some a bit nervous and some more adventurous studied it for quite some time and verified the fact that there was something out of the ordinary that had presented itself out of that painting, of what seemed and appeared out of a supernatural nature which was confirmed by the rest of the media so I put the painting away in my briefcase so as not to get it mixed up in my other works. I had forgotten about it for years except bringing out by visits from a number of sceptics and was years after our explosion at Six Bells. But most of my work was tied up with events that happened at my pit in 1960, until one day a reporter came from the Gwent Gazette and asked to look at the paintings and asked questions like they usually do, and said, 'Don't you understand that the painting that you have done in 1985 was twenty five years after the disaster so it appears that this particular painting was to commemorate the disaster of twenty five years ago'. And funny enough I had finished it at approximately half past ten to ten forty five or near enough at the same time or whereabouts the time of the tragedy had occurred and the reporter said to me 'hat a remarkable coincidence that you had that numbness in your arm'. So photographs were taken of the painting by the Gazette but failed in their attempt to get near the images that appeared, even using different filters in the camera and this is perfectly true and like if it was for me alone, but I was holding an exhibition at Abertillery Comp school which attracted many visitors including a head teacher from Monmouth who was accompanied by a friend who probably was a teacher also and after a while they approached me and said 'Don't you understand Chopper that your soul is in that painting'? They were both convinced that it was so and the only explanation they could find. I remember that occasion very well because the exhibition was opened by the presiding mayor of Blaenau Gwent accompanied by the German Mayor, the exhibition was in aid of the Mayor's appeal for that I cannot remember what appeal it was for, but there was a voluntary donations put into a container, but it was a successful venture, that was before Peter Law became an Assembly member and now a MP. That comment from that head teacher and friend near convinced me that their explanation of me and the painting was true, so I don't know but it makes you think about it but the evidence is plain to see and the proof of the pudding is in the eating and I have the proof of the pudding at my home today as some of the visitors can testify, to the phenomena of the national press and TV interviews have tried to get a true picture from the original but have failed.

During the course of a few years I have had a number of visits from Professor John Harvey of Aberystwyth University School of Aft who was interested in the strange paranormal activities when exhibited the painting at my home and asked him what title I should put on the painting, he said 'The Untitled' and while the wife and I visited my exhibition at Cardiff's National Museum of which I was the main contributor, we sat on benches provided for visitors and on the bench was Catalogue of nine artists that were distributed throughout Galleries and after scanning through the catalogue there was the painting I have described to you and he explained of the fierceness and rapidity in which these images involuntary filled my imagination that persuaded me that they emanate from a supernatural source to which other artists can testify, but I will not go into the story too far, but on Professor Harvey's writing of myself give a true account

Blue Scars

in suggesting that my extraordinary experience of drawing a disaster twenty five years later whether interpreted as an example of the paranormal but testifies and bear testimony to other haunted collieries in South Wales during (1904-1905). So in fact I am not by myself in believing that something out of the ordinary caused the result of my painting and coming from a authority of such as Professor Harvey put my mind at rest, not because of my feelings but to know that others have had the experience of paranormal activities that I have described earlier in my book, and that other collieries can testify to, that is why the TV had cause to visit my home.

But one TV company not connecting with my previous story were visiting the South Wales valleys to find out the feelings of miners and the anger they felt at the closing of their collieries, so they visited two miners that lived in other valleys, had their version and how they felt and quite a surprise to me phoned and asked would an interview be possible and could they come and take photographs of all my work, owing to the fact that the previous interview with the other two had recommended me as an artist of their work. I agreed they would make an appointment as to when they would arrive, the day arrived for the film crew to visit me, I didn't know how many would be arriving, and my wife Tilly made the usual arrangement of providing the catering which she always did. We didn't know how many to expect and when they arrived they were a party of five and they announced they had come from the Birmingham studios, the producer also came from that area. His name was Mike Hatton, the party consisted of a camera man, the equipment carrier, choreographer and producer and two black people, who was the nicest people anyone would wish to meet and what I liked about the coloured chap was when I asked would it be alright to smoke, he said "of course you can, you can have a roll of my tobacco if you want". I said that I had my own that was similar to his and that put me at my ease. So they prepared the tripods for the camera and the setting of the monitors, I began to worry that if these things would interfere with my implant of a pacemaker and the amount of electrical equipment that surrounded me, apparently it did not, so everything went according to plan, the producer wanted to know my feelings about the closure of the pits and I told them in no uncertain terms, I don't know what went on between their last interview but they were only interested in what I was going to say, and my version. I told them if I could have done anything about it I would catch hold of Thatcher and Heseltine with Mr. Greggor and put them down Beynon's pit and seal the top of the pit and when I said that the producer was a bit concerned and hesitated at that point to erase it from the film, but one of the film crew said to him, this is the reason why we are interviewing him to get his reaction, so what can we expect, so they decided to keep the film going and after some more refreshments by my wife asked could they see the Ghost painting as one of the crew called it. When asked where they were previously shown I told them they were shown on a programme called Weird Wales film on BBC and Nicola Heywood film programme High Performance HTV and they said this programme will be called (The pit and the pendulum) (a burning embers production at Birmingham studios), so after another cup of tea they packed up and promised me a video of my interview to which they did and is now still in my possession amongst my other videos of Weird Wales and High Performance, plus my live talk with Roy Noble

Blue Scars

so I am satisfied. In the beginning I had a sense of going somewhere and I have partly achieved total satisfaction, but I will plod on and be thankful to what I have achieved so far. All of this started because of a bit of oil and dust and chalk and imagination and the will, at least it has kept my mind active and my memories alive but now mother nature has put her foot down and had the last say and sealed the pits forever, but I am coming to the conclusion of my story morbid that I may have been in parts, but it's a fact and true account of one man's experience in the dark world of mining as others can testify, even the younger element have been left to find jobs outside the industry and try to find work outside the industry and try to adapt to a new way of employment after pit closures. Many are finding it hard to find different skills because you don't need a sledge hammer in a biscuit factory but at my age like many others have no hope of work, because we are at the later stages of our life, so we are left on the scrap heap as with other workers of another industry the steel workers who suffered the same let down and depend on the mere pittance in the pension scheme, but it's nice to have a grumble sometimes but no one is going to take notice of you, because they have troubles of their own I expect, but this is my safety valve to let out my feelings, by not being active anymore, but when I have a pint I look around me in the pub and I think to myself I ought to be bloody well lucky to be around to have that pint, there are millions that would change places with me, this is why God has given me the opportunity to enjoy the rest of my life and to try and make my ambition come true, to be the caretaker of memories and a trustee of the past, but I wonder if the biography of my life will ever bear fruition, after my effort in compiling this true story of my life down the pits, contribution to a forgotten era and hope that someone somewhere will continue to carry on with my work as an artist and writer, that all I have to say in conclusion of my book.

So, all the best

John (Chopper) Davies
2005

A big thanks to Jill Mercer, Vanessa Smith and Lynda Sage.
Without your efforts I could never have done it.
Thanks again, Elaine.

Lightning Source UK Ltd.
Milton Keynes UK
UKOW052334121011

180245UK00001B/13/P